SpringerBriefs in Regional Science

SpringerBriefs present concise summaries of cutting-edge research and practical applications across a wide spectrum of fields. Featuring compact, authored volumes of 50 to 125 pages, the series covers a range of content from professional to academic.

SpringerBriefs in Regional Science showcase emerging theory, empirical research and practical application, lecture notes and reviews in spatial and regional science from a global author community.

More information about this series at http://www.springer.com/series/10096

Marco Ferretti · Adele Parmentola

The Creation of Local Innovation Systems in Emerging Countries

The Role of Governments, Firms and Universities

Springer

Marco Ferretti
Dipartimento di Studi Aziendali e Quantitativi
Università degli Studi di Napoli Parthenope
Naples
Italy

Adele Parmentola
Dipartimento di Studi Aziendali e Quantitativi
Università degli Studi di Napoli Parthenope
Naples
Italy

ISSN 2192-0427 ISSN 2192-0435 (electronic)
SpringerBriefs in Regional Science
ISBN 978-3-319-10439-3 ISBN 978-3-319-10440-9 (eBook)
DOI 10.1007/978-3-319-10440-9

Library of Congress Control Number: 2015930503

Springer Cham Heidelberg New York Dordrecht London

Printed on acid-free paper

Springer International Publishing AG Switzerland is part of Springer Science+Business Media (www.springer.com)

Contents

Chapter 1
Introduction

Abstract In recent decades the newly industrialised economies have experienced intensive technological learning and made remarkable progress in reducing the gap to the developed countries A large part of the studies address the inquiry into emerging-nations technological development by adopting a micro-level perspective nevertheless, just as in the long-industrialised nations, the development of innovation in single firms is closely tied to the local innovation system in which they are inserted. These peculiarities of LISs in developing countries generate much scepticism among scholars about the appropriateness of the entire LIS concept to such national cases. Nevertheless the specificities of innovation systems in developing countries need not be interpreted as failures to conform to the traditional models of LISs. Instead, they should stimulate a reinterpretation of the model in a dynamic mode. The objective of the present work is to resolve the gap in the literature, identifying the characteristics, and particularly the dynamics of formation for innovation systems in developing economies.

Keywords Innovation · Emerging countries · Systemic approach · LIS · Actors · Dynamic model

In recent decades the newly industrialised economies, such as Korea, Taiwan, Singapore, and more recently China, have experienced intensive technological learning and made remarkable progress in reducing the gap to the developed countries (Kim and Nelson 2000; Lee and Von Tunzelmann 2004). Interest in such transformations has led to the growth of a substantial empirical literature focused on describing and explaining how some developing countries, including the newly industrializing economies, have gained ground on the advanced countries, essentially by assimilating and adapting their more or less obsolete technologies (Vernon 1966; Utterback and Abernathy 1975; Kim 1980; OECD 1992; Dahlman et al. 1985).

A large part of the studies address the inquiry into emerging-nations technological development by adopting a micro-level perspective that compares the

M. Ferretti and A. Parmentola, *The Creation of Local Innovation Systems in Emerging Countries*, SpringerBriefs in Regional Science,
DOI 10.1007/978-3-319-10440-9_1

processes adopted by the latecomer firms to innovation processes in the firms of developed countries. These studies offer only a partial vision of the phenomenon of innovation development in emerging nations. This is because, just as in the long-industrialised nations, the development of innovation in single firms is closely tied to the local innovation system in which they are inserted, and is not the result of individual choices but rather of collaboration among different actors, such as firms, institutions and research agencies (Metcalfe 1995; Bell and Albu 1999; Diez and Berger 2005).

The term "local innovation system" originally arose from the literature on national and regional innovation systems. According to Freeman's (1987) pioneering definition, the national innovation system (NIS) is essentially "a network of institutions in the public and private sectors whose activities and interactions initiate, import, modify and diffuse new technologies".

Throughout the 1990s, scholars advanced numerous theoretical and empirical studies illustrating the significant characteristics of NISs. In particular, Lundvall (1992) examined the features of the learning processes in such systems, which he stressed as developing from institutional roots. Nelson (1993) instead conducted analytical comparisons of the experiences in different NISs, in studies that brought out the complexity and heterogeneity of the cases. Edquist (1997) contributed an identification of the economic and social determinants that permit development of NISs.

However in these same years, several authors began to question the advantages of considering innovation systems from a national perspective and started to move the focal point to regions (Cooke et al. 1997). The region is considered as an entity smaller than the nation, yet still with significant governance capacities over local territories, as well as a level of cohesion that distinguishes it from the overall State and from the other regions (Cooke et al. 1997: 480). Given the apparent functionality of this conception, authors such as Todtling and Kaufmann (1999) argued that it is at this level that analysis of innovation systems would be most fruitful.

Nevertheless in more recent years some studies, departing from the concept of the regional innovation system (RIS), find that the region is still inadequate as the fundamental geographical entity for describing the localised nature of the innovation process. The region can in reality be an entity either too small or too large to host an innovation system.

Some authors propose the enlargement of RIS concept to include the extra-regional relationships that influence the innovative process (Coe and Bunnel 2001). Other studies reveal how the region is in fact, on the contrary, an entity too large to comprehend the phenomenon of the innovation system, which is often localised in more restricted areas (Rantisi 2002). Consequently many scholars have applied the term "local innovation system" to identify a network of technologically specialised and locally situated firms, institutions and research agencies that are involved in a process of collective learning, where this process is not limited to the local geographical borders (de la Mothe and Paquet 1998; Cooke 2001, 2004; Asheim and Coenen 2005).

In recent years, studies concerning LISs have been oriented towards the analysis of the differences and similarities in the systems of the western nations and their regions. In spite of the variations due to territorial contexts, all analyses seem to confirm that systems of innovation feature certain common aspects, among these: (i) the spatial concentration of actors and a localised nature in the innovative processes (innovation originates and develops in the same territory); (ii) the coexistence in the same territory of three types of actors—firms, research agencies and government institutions—that participate equally in the innovative process; (iii) the existence of a close network of relations among actors, which stimulates the type of "learning by interacting" that favours development of innovation.

However, these elements do not seem compatible with the characteristics observed in systems of innovation in developing economies.

First of all, foreign firms and their direct investments play a much more prominent role in innovation systems of developing countries than they do in developed countries (Diez and Berger 2005). In a way, the innovation systems of developing countries are significantly globalised: the majority of new knowledge and technologies is acquired from extra-regional sources via direct investment or technology licensing (Ernst 2002). Further, innovation systems in the developing nations generally seem to appear as forms of "learning systems" rather than "innovation systems", since their fundamental aim consists of supporting processes for the learning and acquisition of pre-existing technological knowledge rather than stimulating new knowledge production (Mathews 2001; Viotti 2002; Lall and Pietrobelli 2005).

Also, developing-economy innovation systems are characterised by the fragmentation of the actors and their linkages (Intarakumnerd et al. 2002). Some actors are still missing or incapable of contributing to innovation activities, and the levels of social embeddedness and trust are still insufficient to support evolution of strong linkages among the actors. Finally, governments play a significant role in building successful innovation systems in developing countries (Yusuf and Stiglitz 2001). Basic investments in the innovation infrastructure are necessary and have to be made by the public sector. The necessary investments include human capital development, academic capacity building, and the promotion of industrial innovation, local content, and technology transfer. The interventionist governments sometimes found in developmental States tend to dominate any innovation systems, which are then centralised in character. Governments in these States then also face the challenge of allowing competitive forces to take effect.

The role of universities as alternative sources of knowledge tends to be marginalised, compared to perceptions of their role in long-industrialised nations. Rather than academic discussion inquiring into science-based industries, research spin-offs, or entrepreneurial universities, as we see in the industrialised economies, it seems that the role of universities in developing countries is truly less central to economic upgrading and technological progress. The contributions of universities are largely limited to education (Liefner and Schiller 2008; Altbach and Salmi 2011).

These peculiarities of LISs in developing countries generate much scepticism among scholars about the appropriateness of the entire LIS concept to such national cases.

Nevertheless some authors affirm that the specificities of innovation systems in developing countries need not be interpreted as failures to conform to the traditional models of LISs. Instead, they should stimulate a reinterpretation of the model in a dynamic mode (Arocena and Sutz 2000; Chaminade and Vang 2008; Lundvall et al. 2009).

The objective of the present work is to resolve the gap in the literature, identifying the characteristics, and particularly the dynamics of formation for innovation systems in developing economies.

In the first chapter of this book we describe the theoretical background, considering the literature on local innovation systems and analysing the studies on technological development of emerging countries. Starting from the theoretical background we then define our approach to describing local innovation systems in emerging economies.

Integrating the various analytical perspectives, we can imagine a sort of ideal path for the constitution of innovation systems, depending on the actors involved and the developmental level of the local entrepreneurial system. In detail, we hypothesise that the traditional configuration of innovation systems, with three typologies of actors (firms, research centres and local institutions) that interact to facilitate the processes of innovation development, tends only to appear in developing nations in the last phase of the innovative process.

During the earlier phases of knowledge acquisition and assimilation we can further hypothesise the existence of other types of systemic structures where a specific actor tends to drive the development of the LIS. In the case that the development of the entrepreneurial system is at a very low level and the actors are insufficient for the creation of an LIS, the only means to engage such creation mechanism lies in strong government intervention. In the case that the local entrepreneurial system is more developed and a large company already exists in situ, it could be that this firm, pushed by the necessity of broadening its competitive horizons, would engage virtuous mechanism for the creation of an LIS. In other cases the engine of the creation of the LIS could be a university, which broadens its mission from that of a research and training institution to that of an actor that guides entrepreneurial development in an area.

The emergence of the different configurations depends on the degree of development of the local entrepreneurial system and foresees the existence of a specific actor capable of engaging the creation process. Thus, according to the actor involved, we distinguish three configurations of LISs:

- the government-driven LIS;
- the firm-driven LIS;
- the university-driven LIS.

In the following chapters of this volume, we present a series of further analyses of the theoretical and empirical literature to develop a more precise understanding of the dynamics of the different configurations of LISs, accompanied by case studies of particular instances for each model.

In Chap. 3, we provide a more detailed consideration of the different approaches that developing-states governments can adopt to build and drive local innovation systems, and then present and compare two case studies of government-driven LISs in Iran and Dubai (United Arab Emirates). In Chap. 4 we analyse the role of large firms becoming the hub of an innovation system, and the different functions that they can implement to drive the system's development. As an example of a firm-driven innovation system we describe the role of Acer Inc. in an LIS centred on the province of Hsinchu, Taiwan. In Chap. 5 we analyse the cases of emerging-nations innovation systems where research assumes a central role. In this chapter we first discuss the different roles that universities can assume to drive local development, and then analyse the role of universities and research institutions in the case of a biotechnology cluster in Singapore.

The volume concludes with a short review of the entire work, including two figures that summarise the evolutionary concept of the LIS and classify the case studies provided, relative to the theoretical approach. We also very briefly suggest some of the specific utilities of the approach presented, as well as general directions for further research.

References

Altbach PG, Salmi J (eds) (2011) The road to academic excellence: the making of world-class research universities. World Bank, Washington, DC

Arocena R, Sutz J (2000) Looking at national systems of innovation from the South. Industry and Innovation 7(1):55–75

Asheim B, Coenen L (2005) Knowledge bases and regional innovation systems: comparing Nordic clusters. Res Policy 34:1173–1190

Bell M, Albu M (1999) Knowledge systems and technological dynamism in industrial clusters in developing countries. World Dev 27(9):1715–1734

Chaminade C, Vang J (2008) Globalisation of knowledge production and regional innovation policy: supporting specialized hubs in the Bangalore software industry. Res Policy 37:1684–1696

Coe NM, Bunnell TG (2001) Spaces and scales of innovation. Prog Hum Geogr 25(4):569–589

Cooke P (2001) Regional innovation systems, clusters, and the knowledge economy. Ind Corp Change 10(4):945–974

Cooke P (2004) Special issue: globalisation of biotechnology: Editorial: the accelerating evolution of biotechnology clusters. Eur Plan Stud 12(7):915–920

Cooke P, Uranga MG, Etxebarria G (1997) Regional innovation systems: institutional and organisational dimensions. Res Policy 26:475–491

Dahlman C, Westphal LE, Kim L (1985) Reflections on acquisition of technological capability. In: Rosenberg N, Frischtak C (eds) International technology transfer: concepts. Measures and Comparisons, New York, Pagamon

Diez JR, Berger M (2005) The role of multinational corporations in metropolitan innovation systems: empirical evidence from Europe and Southeast Asia. Environment and Planning A 37(10):1813

Edquist C (1997) Systems of innovation: technologies, institutions and organizations. Pinter, London

Ernst D (2002) Global production networks and the changing geography of innovation systems: implications for developing countries. Econ Innovation New Tech 11(6):497–523

Freeman C (1987) Technology and economic performance: lessons from Japan. Pinter, London
Intarakumnerd P, Chairatana P, Tangchitpiboon T (2002) National innovation systems in less successful developing countries: the case of Thailand. Res Policy 31:1445–1457
Kim L (1980) Stages of development of industrial technology in a developing country: a model. Res Policy 9:254–277
Kim L, Nelson RR (2000) Technology, learning and innovation: experiences of newly industrializing economies. Cambridge University Press, Cambridge
Lall S, Pietrobelli C (2005) National technology systems in sub-Saharan Africa. International Journal of Technology and Globalisation 1(3–4):311–342
Lee T, Von Tunzelmann N (2004) A dynamic analytic approach to national innovation systems: the IC industry in Taiwan. Res Policy 34(4):425–440
Liefner I, Schiller D (2008) Academic capabilities in developing countries—a conceptual framework with empirical illustrations from Thailand. Res Policy 37(1):276–293
Lundvall B (1992) Introduction. In: Lundvall B (ed) National systems of innovation: towards a theory of innovation and interactive learning. Pinter, London
Lundvall BA, Joseph KJ, Chaminade C, Vang J (eds) (2009) Handbook of innovation systems and developing countries: building domestic capabilities in a global setting. Edward Elgar, Cheltenham
Mathews JA (2001) National systems of economic learning: the case of technology diffusion management in East Asia. Int J Technol Manage 22(5–6):455–479
Metcalfe S (1995) The economic foundations of technology policy: equilibrium and evolutionary perspectives. Handbook of the economics of innovation and technological change, p 446
de la Mothe J, Paquet G (eds) (1998) Local and regional systems of innovation. Kluwer, Dordrecht
Nelson R (1993) National systems of innovation: a comparative study. Oxford University Press, New York
OECD (1992) Technology and economy: the key relationships. OECD, Paris
Rantisi NM (2002) The local innovation system as a source of 'variety': openness and adaptability in New York City's garment district. Reg Stud 36(6):587–602
Todtling F, Kaufmann A (1999) Innovation systems in regions in Europe: a comparative perspective. Eur Plan Stud 7:699–717
Utterback JM, Abernathy WJ (1975) A dynamic model of process and product innovation. Omega 3(6):640–656
Vernon R (1966) International investment and international trade in the product cycle. Q J Econ 80:190–207
Viotti E (2002) National learning systems: a new approach on technical change in late industrializing economies and evidences from the cases of Brazil and South Korea. Technol Forecast Soc Chang 69(7):653–680
Yusuf S, Stiglitz J (2001) Rethinking the East Asian miracle. Oxford University Press, Oxford (UK)

Chapter 2
Local Innovation Systems in Emerging Countries

Abstract Rather than being a spontaneous phenomenon, the creation of local innovation systems (LISs) in emerging nations typically involves a process driven by an individual actor. The actions of some lead organisation bring about the involvement of other ones, with mechanisms of linkage between them. Empirical evidence shows different types of lead actors, including cases of LIS formation driven by governments, by firms, and by universities or research centres.

Keywords LIS · Emerging nations · Governments · Firms · Universities · Dynamic approach

2.1 The Theoretical Basis for the LIS Concept

The concept of innovation has undergone profound changes since its first appearance in the literature of the early 20th century. In recent years there has been decisive shift away from an essentially linear conception of the development of innovation towards ideas of interactive and systemic processes, which involve not only the different functions of the firm itself (Kline and Rosenberg 1986), but also its relations with other organisations.

In fact, innovation does not originate in any single organisation, but as the result of a systemic process (Fagerberg 2004). The firm collaborates with other actors, such as additional firms, research institutions and government agencies, both in the development stages and in the implementation of innovation. Government usually assumes a particularly important role, since the implementation of State policies to incentivise or block the processes of innovation inevitably conditions the results, often to a remarkable extent.

The recognition that government institutions play an important role, and that interactions in the development process can take place more easily if the actors are in geographic proximity, has underlined the relevance of the territorial variable in stimulating innovation (Lundvall and Johnson 1994; Etzkowitz and Leyedsdorff 2000).

M. Ferretti and A. Parmentola, *The Creation of Local Innovation Systems in Emerging Countries*, SpringerBriefs in Regional Science,
DOI 10.1007/978-3-319-10440-9_2

Studies on the role of the territorial variable in firm interaction and in innovation development descend from the original Marshallian literature on the industrial district (Marshall 1930), which eventually gave rise to studies concerning both economic geography (Asheim 2000; Feldman 1994, 1999, 2000, 2001; Gertler 2003; Clark et al. 2000; Howells 1999, 2002) and what are called "national innovation systems", or NISs. This latter current of study integrates numerous contributions from quite disparate fronts, including the works on "learning regions" (Asheim 1996), on the "innovative milieu" (Crevoisier 2004; Camagni 1995), and the triple helix model (Etzkowitz and Leydesdorff 1997; Leydesdorff 2000), as well as on business clusters (Porter 1990, 1998; Swann and Prevezer 1996; Baptista and Swann 1999).

According to Freeman (1987), the national innovation system is essentially "a network of institutions in the public and private sectors whose activities and interactions initiate, import, modify and diffuse new technologies".

Throughout the 1990s, scholars advanced numerous theoretical and empirical studies illustrating important characteristics of NISs.[1] In particular, Lundvall (1992) examined the features of the learning process in such systems, which he stressed as developing from institutional roots. Nelson (1993) instead took an approach of comparing the experiences of different NISs, in studies that brought out the complexity and heterogeneity of the different cases. Edquist (1997) contributed an identification of the economic and social determinants that permit development of NISs.

However in these same years, several authors began to question the advantages of considering innovation at the national level and in fact departed from the NIS concept, substituting it with analyses of the regional system (Cooke et al. 1997). The region is considered as an entity smaller than the nation, yet still with significant governance capacities over local territories, as well as a level of cohesion that distinguishes it from the overall state and from the other regions (Cooke et al. 1997: 480). Observing the functionality of this conception, authors such as Todtling and Kaufmann (1999) argue that it is at this level that analysis of innovation systems will be more fruitful.

According to some authors the systemic interactions among actors at local levels are intrinsically unique, and thus hard to measure and compare across different systems. A potential solution to this problem is the "evolutionary integrated" view of regional innovation systems (Iammarino 2005). Under the broader theory of evolutionary economics, an effective analysis of innovation processes can be developed by studying the mid-ground between the macro-level of national systems and the micro-level of the individual actors. This meso-economic approach serves to

[1] Studies of national innovation systems can be roughly divided into "broad" and "narrow" approaches. The narrow approach considers NISs as composed only of firms and research institutions, such as universities and R&D centres, as well as local government. This conception coincides with the "triple helix" model (Etzkowitz and Leydesdorff 2000). The broader approach considers that the elements composing the NIS also include all the economic-structural and institutional-contextual aspects that influence the research and business spheres (Asheim and Isaksen 1997).

describe "the essential thing that is changing in a process of evolutionary economic change" (Dopfer et al. 2004, p. 269) and can thus account for local and regional variety in terms of absorption, diffusion and generation of new knowledge. A number of fundamental concepts, such as those of industrial districts, learning regions, innovation systems and others, can all be referred back to this "meso" perspective, under the general evolutionary economics point of view. Further, an integrated micro-meso-macro approach to the socio-institutional determinants of innovation performance offers a means of dealing with the heterogeneity and path dependency observed in empirical studies of regional and local economies, particularly in terms of "local structural regularities from past knowledge accumulation and learning" (Iammarino 2005, p. 503). These are in turn the factors that shape and constrain the innovation processes and growth opportunities.

Thus in recent years the concept of the regional innovation system, or RIS, has gained ever more ground. The RIS is defined as a social system that features systematic relationships between different groups of actors, from both private and public sectors, for purposes of increasing and improving the learning capacities localised in a particular region (Doloreux 2002).

However certain studies examining the definition of RIS identify that the region could be a geographic entity too small or alternatively too large to host an innovation system. Some authors propose an enlargement of the concept of the regional innovation system to also include the extra-regional relationships that influence the innovative process (Coe and Bunnel 2001). Other studies instead illustrate that the region is an entity too large for the consideration of innovation systems that are in fact often localised in a more restricted area (Rantisi 2002).

To bypass the problem of the spatial configuration of innovation systems, Oinas and Malecki (2002) propose the concept of a spatial innovation systems (SISs), which consist of "overlapping and interlinked national, regional and sectoral systems of innovation which all are manifested in different configurations in space". In introducing this concept, the authors suggest that innovation systems are worked out differently in space and can exhibit different spatial configurations. They may originate in one place, but often spread beyond local, regional, and even national borders. Technological evolution then occurs through the interplay between elements of national, subnational, and transnational innovation systems, which produce flows of innovation and have different inherent capabilities in keeping up with state-of-the-art practices on the various technological frontiers.

Moulaert and Sekia (2003), in a critical review of studies that explain the localised nature of the innovative process, illustrate that the different conceptions of systems are often interrelated. Apart from the differences in territorial contexts, all the analyses seem to confirm the existence of certain common aspects. From these results we identify any innovation system as: (i) the spatial concentration of specific actors and the localised nature of the innovation process, such that innovation originates and develops in the same territory; (ii) the coexistence in the single territory of firms, research and government institutions, representing the three types of actors that participate equally in the innovation process; (iii) the existence of inter-meshed relations between the actors, which stimulate a process of learning by interacting and thus favour the development of innovation.

Local and regional innovation systems are spatial configurations that emerge as covenants of identity, and as frameworks for belonging and learning. Thus what contributes to making an NIS, RIS or LIS is not only the aspect of geographical concentration but also the process of collective learning (de la Mothe and Paquet 1998).

Consequently many scholars use the term "local innovation system" to identify a place where a process of collective learning is localised, but where this process is not necessarily limited to occurring within the local geographical borders. The main function of LISs is to generate new, practical knowledge and to commercialise it. The local economy gains evolutionary momentum through the generation of innovation, produced by recombining various types of endogenous knowledge with externally sourced new knowledge. The interaction of the local actors with the external environment determines the extent to which the economy is subject to positive or negative lock-in, and the creation of new paths of development. Thus the degree to which the actors in a local economy can access, understand and convert leading-edge knowledge to new products and services determines their capacities to generate new pathways and renew old ones. On the other hand, where the local actors are far behind the new knowledge frontiers, there will be an increasing likelihood of the economy becoming locked into ageing technologies and going into decline (Martin and Simmie 2008).

2.2 Local Innovation Systems: The Search for a Definition

The literature offers various conceptual definitions of the local innovation system. Cooke et al. (1997) define the LIS as a system "in which firms and other organisations are systematically engaged in interactive learning through an institutional milieu characterised by embeddedness". Asheim and Isaksen (1997) add that "a [local] innovation system consists of a production structure (techno–economic structures) and an institutional infrastructure (political–institutional structures)". Doloreux (2002) instead concentrates on the aspect of the LIS as a social system, which features relations among different groups of public and private actors, acting in a systematic manner and resulting in increased learning potentials within the given region.

Various authors have also proposed criteria that identify and define the component elements of an LIS. Todtling and Kaufmann (1999) stress that the central elements are the firms that belong to the region's principle industrial sectors, flanked by those that operate in complementary sectors. These firms compose different types of networks that operate both within the region and beyond its boundaries, based on supplier-client relationships, cooperation and information exchange. The relational networks created are the vehicles for knowledge and information exchange, and thus enrich the territorial knowledge base.

In addition to the firms, LISs include research institutions and universities, which serve as knowledge creators and training centres. However these will create wealth in the territory only if the knowledge generated is actually transferred to the local firms.[2]

Another important aspect influencing the local system's capacities is the quality of the workforce. This component is understood not only as the individuals engaged in research, but also the entire spectrum of personnel in production, marketing and human resources management. From this perspective, activities in education and training represent another important factor in the LIS.

In addition to the fundamental elements of the firms, workforce, and research and educational organisations, financial institutions must also play a role. These assist the firms to take on the substantial investments involved in innovation processes, and thus activate their projects. Still further components are the industrial associations, and the agents roughly grouped as "business innovations centres". These are the hubs of scientific research and the centres for scientific and technological transfer, organised in different manners and with different specific targets, but with the consistent aim of supporting firms in the general goal of overcoming barriers to innovation.

Much as in the above scheme advanced by Todtling and Kaufmann (1999), Doloreux (2002) identified the component actors of the LIS as the firms, government institutions, research and training institutes that interact and receive support from the regional policies intended to promote innovation processes. Cooke (2001) emphasises that what distinguishes a local innovation system is not so much the firms and educational institutions that compose it, but rather their capacity to interact for the achievement of a sort of collective learning, similar to what occurs in Lundvall's (1992) "learning region".

According to Martin and Simmie (2008), local innovation systems have the following characteristics: (i) sectorally and institutionally diverse knowledge-generating businesses and institutions, able to draw innovative ideas from many potential sources; (ii) firms at high levels of specialisation, capable of supplying the best to national and international markets; (iii) commercial and marketing know-how, based on awareness of international markets and technological conditions; (iv) a social culture that demonstrates and tolerates diversity, and offers new ideas and ways of doing things; (v) firms that are able to exploit knowledge and to support knowledge applications by others; (vi) high levels of technical sophistication among producers and users of technology; (vii) economies of scale; (viii) international knowledge spillovers from sophisticated customers, including locally-represented multinational companies, which provide the LIS with information on leading-edge knowledge, products and services. While all these characteristics are objectively possible elements of an LIS, individual localities will have different

[2] This very important consideration establishes that even in areas with advanced research centres, if there is no dialogue with local enterprise and such a situation continues, then the territory will not develop a local innovation system.

combinations and strengths in each of them. In many cases one or more of these phenomena will be missing altogether.

Some scholars have given definitions of the LIS that are much broader, however certain of these risk a confusing level of overlap with other well-known concepts, such as the industrial district,[3] science parks and technopoles, the innovative milieu, or the learning region (Doloreux and Parto 2004).

Asheim (1998) attempted to systemise the various contributions and propose a definition of the LIS that would reflect the conceptual variation and the empirically observed multiplicity of relationships seen to exist in regional productive and institutional structures. He distinguished three types of system (see also Asheim and Isaksen 2002; Asheim et al. 2003; Asheim and Coenen 2005): (i) the territorially embedded local innovation system; (ii) the regional networked local innovation system, and (iii) the regionalised national innovation system.

The territorially embedded system type is characterised by an interactive network of firms that base their innovation activity on processes of reciprocal learning, stimulated by geographic and relational proximity, but in absence of direct links with government institutions and research institutes. The best examples of territorially embedded LISs are the networks of small-medium enterprises found in industrial districts.

The second type of LIS, the regionally networked system, is again characterised by the presence of a network of firms and organisations embedded in a specific region, which interact and generate reciprocal learning. However in this case government intervenes to develop more systematic networking through the strengthening of regional institutional infrastructure. There is also a greater role on the part of regional R&D laboratories, as well as structures and organisations for professional education, and other local organisations involved in various aspects of the firms' innovation processes. These systems have a more planned, systemic character, originating in constant public-private cooperation. Various authors consider the regionally networked system, as originally observed in Germany, Austria and the Nordic nations, as the ideal type of LIS.

The regionalised national innovation system differs from the first two LIS types in two aspects. First, there is greater functional integration between the firms and the national or international infrastructure of the innovation system. In this type, innovation activity develops primarily through cooperation with actors outside the region. In this development model the exogenous actors and inter-systemic relationships thus play more important roles. A second difference is that cooperation between organisations within such systems less often involves incremental types of innovation, and instead fits more closely with linear models of innovation, meaning innovation that takes place primarily around specific projects capable of developing radical innovations, which in turn ultimately proceed from formalised scientific and analytical knowledge. Specific examples of regionalised national innovation

[3] Hommen and Doloreux (2003) in fact argue that Marshall's historic study on industrial districts can be viewed as the earliest conceptual approach concerning local innovation systems.

systems include technological and science parks or technopoles, composed of R&D laboratories of large companies, at times flanked by government agencies, universities and technical institutes localised in very close proximity, but where all these actors generally present weak links to local firms.

The above tripartite classification system, initially advanced by Asheim, is similar to one proposed by Braczyk et al. (1998). In their classification, the three typologies of LIS are "grassroots", "network" and "dirigiste", distinguished on the basis of their governance styles and the modes of enacting the technology transfer processes.

The "grassroots" model describes a region where technology transfer is primarily organised and promoted at the local level, within the region's own organisations and government structures. The "network" model is the result of interaction of local, national and global-level institutions in the governance of technology transfer at different levels. Finally the "dirigiste" model envisions a situation where the governance of technology transfer is essentially centralised under national institutions. France is cited as the type example of the "dirigiste" model.

Howells (1999) offers a definition of national innovation systems that distinguishes two typologies of LIS, in a somewhat similar sense to the preceding tripartite model. In this conception, the classification criteria are "top-down" or "bottom-up", depending on the connections between the local and national systems. The top-down LIS takes form as a subsystem of the national system, stimulated by national policies at the local level. In the bottom-up case, the local systems originate autonomously and it is their independent interaction that ultimately leads to the development of the national innovation system. The bottom-up LIS presents three specific elements: a local structure of autonomous governance, strong local industrial specialisation, and high capacities for innovation among local firms.

2.2.1 Towards a Narrow Definition of the Local Innovation System

The numerous definitions and classifications of LISs, as well as the overlap with other important concepts such as the industrial district, learning region, innovative milieu and economic cluster (Asheim et al. 2011), risk confusion for the practical application and usability of the entire concept.

In order to understand how a local system can develop in a specific geographic area it seems more useful to draw on a narrow definition of LIS that permits immediate comprehension of its specificities and constituent elements.

For this, we identify four elements as defining the existence of a local innovation system:

- a network of innovative firms, localised in the same area and bound by horizontal and vertical relationships;
- a set of research and educational institutions, such as universities and research centres, which generate scientific knowledge that contributes to innovative processes;

- a series of infrastructure provisions that incentivise the localisation of innovative firms within the given area;
- the presence of cooperation mechanisms among all these actors, capable of stimulating reciprocal learning and thereby processes of innovation.

The last of these four aspects is particularly important. It is the cooperation between firms and research institutions that favours not only the exploration and development of the existing knowledge base, but also experimentation in new knowledge through "learning by interaction" (Noteboom 2001). The interactive learning process is favoured by the existence of social interaction mechanisms, which develop readily among actors in close cultural and geographic proximity (Antonelli 1999). Thus the level of social embedding of the various actors represents an important element in the birth of an LIS.

However the identification of social embedding as a fundamental variable still does not resolve the problems of the origins or definition of local innovation systems. For example if we look at industrial districts, there are numerous cases of groups of socially embedded firms, with interaction between them that clearly serves in improving the efficiency and effectiveness of the production process, but where the interaction still does not generate significant innovation. In fact in these systems, the participation of research institutions and universities is very limited or lacking. The nature of the knowledge that is produced and applied in fact does not require modification of any base paradigms. Instead the innovation that arises is an incremental type, which does not permit the kind of break-through in technical frontiers that should represent the true objective of local innovation systems.

Given these considerations, and to distinguish the LIS concept from other types of local systems, we apply two variables indicating the level of social embedding (high or low) and the character of the knowledge base (synthetic or analytic).

2.2.1.1 Level of Social Embeddedness

The concept of "social embeddedness" was originally applied in the study and understanding of personal relationships, more so than economic ones. In the former case, embedding permits easier management of personal relations, through encouraging trust and discouraging development of opportunistic behaviour (Granovetter 1985). Social embeddedness also contributes to trust between firms, and the two variables are positively and very closely related.

Although Granovetter's (1985) original application to economic contexts did not specify the factors that determine embeddedness, the phenomenon is clearly influenced by cultural and geographic proximity between firms (Lyons 2000). Increased proximities favour social interaction, and consequently stimulate the process of knowledge transfer.

In this sense, the phenomenon of collective learning or learning through networking takes place more easily when proximity and levels of social embeddedness are high.

The aspect of social embedding distinguishes the LIS from other territorial concentrations where interaction is based only on market relations. In the case of primarily market types of relations, interaction requires only passive rather than active involvement of the group members (Oughton and Whittam 1997).

2.2.1.2 Type of Knowledge Base

This variable indicates the type of knowledge that circulates within the innovation system, and is thus linked to the type of industrial sector or sectors where the system operates, as well as the innovativeness of the actors within it. The dichotomy between synthetic and analytic knowledge serves as a useful classification for understanding the local system's knowledge base (Asheim and Gertler 2004; Asheim et al. 2011).

Analytic knowledge is fundamental in industries that are heavily based on scientific research, where knowledge is created through rational cognitive processes and formal models. The innovation strategy in such industries is based on access to explicit codified knowledge ("science, technology and innovation", or STI-mode). This is a science push/supply-driven high-technology strategy, able to produce radical innovations. Basic and applied research, as wells as systematic development of new products and processes, are all important activities in these industries. The firms belonging to such systems usually have their own R&D departments, however it is not uncommon that they also draw on research results from universities and other institutions. In systems featuring an analytic knowledge base, industry-university relationships are of fundamental importance for innovation development.

A synthetic knowledge basis is typically found in industrial contexts where innovation takes place primarily through recombination or reapplication of existing knowledge. The process can be defined as "doing, using and interacting", or DUI-mode. It relies on informal processes of learning and development of experience-based know-how. The DUI mode is a user or market-driven model based more on competence building and organizational innovations. It produces mostly incremental innovations. R&D activity is primarily concentrated on physical improvement of the product or production process. Industry-university links exist but primarily concern training activities.

A narrow definition of LIS implies the development of the analytic knowledge type (Asheim et al. 2011).

Combining the dimensions just illustrated, and on the basis of the existing literature, we can identify four typologies of local systems (Fig. 2.1). The typologies are described in detail below.

Cluster (*Quadrant A*): This quadrant represents a class of local systems characterised by low levels of social embeddedness between actors and by a synthetic knowledge base. In this case the situation is not one of a true local system, but simply a group of firms that operate in the same geographic area and that have relationships exclusively of the supplier-client type. This description is coherent

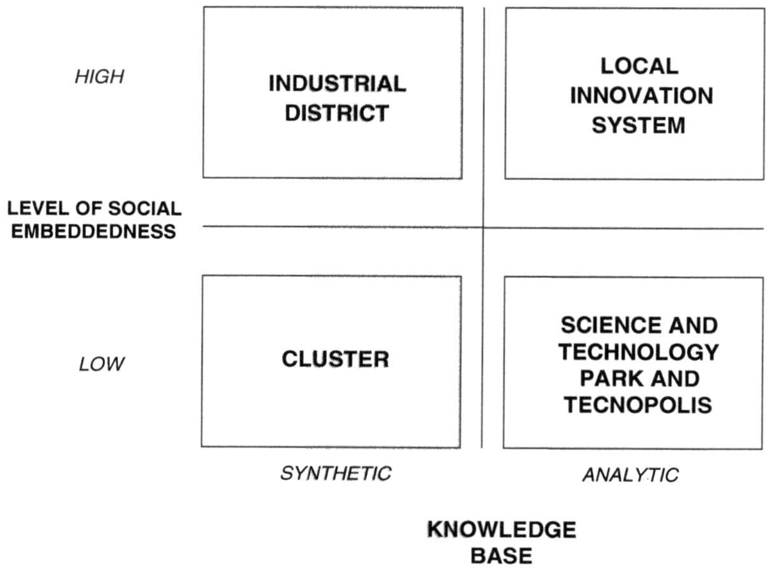

Fig. 2.1 Definition of local innovation systems. *Source* Ferretti and Parmentola (2007)

with Porter's (1998) description of the industrial cluster, understood as a geographic concentration of firms and institutions. Industrial clusters are characterised by the existence of a vast set of firms, connected in a network of primarily supply-type relationships that do not implicate true knowledge exchange between the actors. Porter's definition does not imply the existence of any trust-based relationship between the companies, which would signal profound social embeddedness, or the presence of any explicit attribution of importance to the links between firms and research institutions. Such links would represent the precondition to realising a system with an analytic knowledge base.[4] In fact in conveying the concept of industrial cluster, Porter (1998) cited the example of Californian wine producers. This cluster included 680 winemaking firms and many independent grape growers, as well as a large number of firms belonging to sectors that support both the grape-growing and wine-producing processes. The relationships between firms were of the supplier-client type and did not involve higher levels of knowledge exchange, nor were there relations with local universities.

Industrial district (*Quadrant B*): This quadrant represents a class of local systems characterised by a high level of social embeddedness between actors and a synthetic knowledge base. These features are typical of the systems of Italian firms traditionally defined as "industrial districts". The classical definition of industrial district is a concentration of firms in which it is difficult to distinguish

[4] In literature the term "cluster" is often used as a synonym of local system, with both terms applied in describing different phenomena such as productive cluster, technological cluster, etc. In this study the term "cluster" is used in the stricter meaning as originally defined by Porter (1998).

the boundaries between the community of companies and that of individuals (Becattini 1989), and where economic success is strongly dependent on the socio-cultural context. The existence of reciprocal trust and an "industrial atmosphere" are necessary components in the definition of an industrial district. Geographic and cultural proximity ease the personal relationship between firms and increase the level of social embeddedness. Although some studies (Saxenian 1994) in fact use the expression "industrial district" to describe various local systems that operate in high-tech sectors characterised by an analytic knowledge base, such as Silicon Valley, the classical definition of industrial district does not include the presence of research institutions. Such presence would be necessary for firms to develop the relationships needed for creating an analytic knowledge base. In fact it is only necessary to identify the occurrence of two elements to define a local system as an industrial district: the presence of a community of firms localised in the same area, and the "industrial atmosphere". These considerations are confirmed by the analysis of Italian industrial districts, which demonstrate innovation development of incremental character and scarce relationships with research institutions.

Scientific and technology park, technopole (*Quadrant C*): This quadrant groups a class of local systems characterised by a low level of social embeddedness between actors and an analytic knowledge base. These characteristics are typical of the so-called "applied-science research hubs": proprietary ventures engaged in the development and management of a series of physical or service infrastructures that support the embedding of firms and research institutions in a specific territory, such that these inter-relate and favour economic development in the area (Shearmur and Doloreux 2000: 1066).[5]

The "technopole" is a concept similar to the science and technology park, which arose in the late 1970s following the initiation of the "Technopolis" programme in Japan and the Technopôle policy in France. A technopole consists of a physical and social infrastructure intended to support industrial innovation activity and technological transfer, for purposes of favouring institutional collaboration between universities and industry (Shearmur and Doloreux 2000). The science park and technopole definitions highlight the role of the university and the university-firm pairing as fundamental requisites for the birth of systems based on analytic knowledge. The inter-firm and firm-institution relationships are highly formalised and lack the social embedding that would favour birth of learning by interacting (there is physical proximity among the actors, but not cultural).

[5] According to Shearmur and Doloreux (2000, p. 1066), the full definition of a scientific or technological park is: "a property-based venture which has: existing or planned land and buildings designed for private and public research and development facilities; technology and science based companies relating to support services; a contractual and/or operational relationship with a university or other institution of higher education; a role in promoting research and development by the university in partnership with industry, assisting in the growth of new ventures, and promoting economic development; a role in aiding the transfer of technology and business skills between the university and industry tenants".

Local innovation system (*Quadrant D*): This quadrant represents the class of local systems characterised by a high level of social embeddedness between actors and by an analytic knowledge base. The high level of social embeddedness supports mechanisms of social interaction, which stimulate processes of knowledge transfer. The existence of an analytic knowledge base implies the presence of educational institutions and a strong link between firms and universities.

With the meaning of the local innovation system clarified, it is now necessary to analyse its components in detail, in order to understand the potential features of LISs in emerging nations.

2.3 Types of Actors in a LIS

To identify the actors in an innovation system we must first begin with an understanding of the characteristics of "innovation". According to Giacometti (1987), innovation "… involves processes of discovery and creation, fed by both universal and public scientific input and by forms of specialised and tacit knowledge that are generated within the innovating institution."

This conception implies that it would be impossible to assign firms the entire responsibility for the innovation process. In fact, a firm would be able to produce micro-innovation through independent action, but for structural reasons, in the absence of relationships, it cannot create profound changes in the technological paradigm. Indeed, size represents a critical variable for the production of innovative technology. In this case, size is to be understood both in terms of productive and financial capacity. Particularly relevant are the possibilities for substantial capital investment in a risky process, and the availability of competencies covering the entire innovation process.

In the context of innovation, the firm thus becomes a link in an enlarged cognitive system[6] of a social type (Rowe 1987). The systems tend towards network forms, which involve both profit and non-profit organisations, and above all link the firm to the local system that furnishes the resources essential to firm performance (entrepreneurialism, stimuli for innovation, channels for knowledge circulation, "trust" relationships). The knowledge produced by the firm is added to the other knowledge produced in the network, to which the firm has access, and to that produced in the local environment, where the firm is embedded.

---

[6] In some very recent literature, the process of value creation refers to the firm's capacity to grow its own knowledge heritage. Even in this case the process still cannot be activated by the firm in an isolated, autonomous manner. Rather it must be inserted in a broader system of social relations that link the firm with its environment. From this point of view, the firm's capacity to generate value does not depend on the mere governance of internal relations, but also and above all on the coordination of external relationships, with the objective of maximising the process of knowledge generation.

From this, we can derive the identification of the three types of macro-actors found in all innovation systems. Each type represents different potential sources of knowledge. Any one actor or type alone would settle into low-entropy processes,[7] destined to close on themselves, while in a systemic approach they can succeed in orienting towards synergic production of knowledge.

The actors are:

- firms;
- non-profit research organisations;
- local institutions.

The first actor in the innovation system is the firm,[8] in its most generic form. The firm moves forward on its own path of value creation through processes of product innovation and of seeking sustainable competitive advantage.

The task of a firm engaged in an innovation process is obviously not to analyse the innovation choices of other individual firms, but to understand the links between their innovation path and other specific and highly contextualised variables.

Rendering this difficult are the facts of the flexibility in sectoral boundaries, of research processes that increasingly surpass the sizes of innovative firms, and of circularity in knowledge, all of which tend to decontextualise the innovating company. The object of analysis is thus no longer the firm per se, but its decision to innovate, to share in an innovation path, to look to the long-term benefits that can derive from the process.

Alongside the firm in the innovation-creating system there is a second actor, namely the non-profit research organisation. Here we refer to universities and research centres devoted to basic research, often managed in a manner independent of the field of potential application.

In the innovation process, public institutions carry out a coordinating function in the realm of national and local technology policies, as well as in diffusing technology through public research centres, and in purchasing innovative goods through government demand. The aspect of public education and training systems plays a fundamental role in technological capacity.[9]

---

[7] Economists measure the degree of dispersion of knowledge, or chaos, as the relation between "energy used" (i.e. knowledge actually used) to potential energy, with the value of this ratio in a range between 0 and 1. In measurement, high degrees of entropy indicate high degrees of use (concentration) of energy (knowledge), and indicate risks of entropic death, if not countered by policies for knowledge exchange that lower the entropic index (increase the potential energy). On the other hand, cases of high levels of entropy indicate a high degree of usable energy, and a situation where there is a lack of structural potential to actually use the energy (knowledge).

[8] The distinction between invention and innovation includes the concept that the former stops at the research stage, while the latter is subject to testing and selection in context, and ultimately to diffusion (Schumpeter 1971). Given this, the firm clearly plays an implicit role in the process of creation and diffusion of the invention.

[9] "The character and effectiveness of a nation's system of schooling, training, and retraining, determines not only the supply of skills, from engineer to machine operator, but also influences the attitudes of workers towards technical advances" (Camagni 1992).

Universities in particular play an important role in educating human capital that is capable of generating basic knowledge relevant to technological advancement in industry. Public research organisations contribute to basic research, and in some case also to applied research. Research associations and private consultants play a significant bridging role between the basic science of universities and research centres and the applied research of the firms (Rullani 1989). Universities can contribute directly to local innovation processes in multiple ways. The possibilities are not limited to the patenting and licensing of discoveries made in university laboratories. In addition, universities can help to attract new knowledge resources from elsewhere; adapt knowledge originating elsewhere to local conditions; help to integrate previously separate areas of technological activity in the region and to unlock and redirect knowledge that is already present in the region but not being put to productive use. Most of these university contributions presuppose the presence of local industry (Goddard et al. 2007; Goddard 2011).

However in many innovation systems the role of universities has added remarkably different directions (Lester 2005). Recently, the most innovative strategies for universities and research centres have moved from the model of the closed university to that of forms that are highly attentive to enterprise situations. This changes the objectives and management of university structures, orienting them towards continuous and constant connection with the market. With this type of university policy, now taking hold in most of the world's major innovation systems, the university assumes the role of the central innovator, promoter and incubator of innovative ideas, transforming the approach from pure invention into innovation and from big science to small science.

A similar path is also activated in the case of non-profit research centres, when the invention they produce features characteristics of economic productivity, meaning costs of production and management that permit its broader diffusion in the system of firms. Once again, these inventions are transformed into innovation.

The last actor in the innovation system is represented by the substrate of local institutions, in which the system embeds. This can be summarised, in broad lines, as:

- the scientific and technological infrastructure;
- institutional and socio-political characteristics (e.g. antitrust regulation);
- the characteristics of the labour market;
- the financial system;
- the size and sophistication of demand in the internal market.

However studies have also shown that other institutional factors specific to each nation and geographic area have a determining influence on the rates and procedures of technological progress.[10] Interest is therefore focused not only on

[10] Becattini (1989) defined the entirety of these country or region-specific variables as the national (or local) innovation system. His intention was to propose a concept analogous to that of the "technological regime", which could synthesise those factors that influence patterns of innovation at the geographic level. He thus defined the national (or local) innovation system as "a set of institutional actors that, together, play the major role in influencing innovative performance".

the behaviour of firms on the technological frontier, and not only on the institutions that carry out the most advanced scientific research, but is enlarged to consider all the factors that influence a nation's or region's technological capacity.

In brief, the national innovation system results as composed not only of firms, even if these are the principle actors, but also of other actors and institutions that carry out a relevant role in generating and diffusing technology. Among these are the universities, the public research institutions, research associations and private consultants.

2.3.1 The Relationships Between the Actors and Their Effects in Innovation

Defining the nature of the relationships between the actors in the system implicitly requires the identification of their individual objectives. These inevitably condition the management of the relations and the roles of the partners. Roles and objectives change significantly according to the pairings involved in the relations: university-firm; firm-local environment; university-local environment.

University-firm relations are the most interesting and fruitful in terms of generating innovation. Historically, university-firm relations have followed a continuous growth trend ever since the birth of the first science-based companies in the years after World War II.

Modern cooperation relationships between university and business systems have significant influence on the paths for the creation and diffusion of innovation, from which both actors can gain benefits, although in different manner.

Wherever the paths have been simplified in their mechanisms and reduced in the times required, then the innovation process has resulted as faster and more fertile. Such simplification and easing of times emerges through closer forms of cooperation between the actors, through establishment of spin-off processes, or the presence of further institutions in support of cooperation (e.g. science and technology parks) (Rothaermel et al. 2007).

Silicon Valley is a well-known case where the universities' specific relational capacities, directed at promoting and financing entrepreneurial initiatives, progressively created a relational network where the universities have been both cause and effect of strongly innovative processes.

Where instead the university-firm relations have remained one-sided and non-systematic, or where "artificial" connections develop between actors that have no subjective, structural or contextual motivations for relations (e.g. due only to cooperation incentives), the relationships have not yielded innovation.

Firm-environment relations have also been shown to be fruitful for innovation. Here we again view the environment as the system of social and economic relations that link the firm to the place, as discussed in the previous subsection. The entire literature on industrial districts, and more recently on technological parks, illustrates the reciprocal stimulus and hindrance effects from the relationship that the firm develops with its environment. In fact the processes of creating value through innovation produce positive effects both for the firms and the socio-economic systems that host them.

The observation of this fact has stimulated growing interest in local contexts for the processes of industrial concentration. With such interest in mind, local administrations tend to enhance whatever spontaneous localization of industry might occur by drawing on local policy actions for further stimulus, such as with the offer of serviced properties and areas, technical services and infrastructure.

If we adopt a still broader understanding of "environment" as involving not only the firm and its local place, but also the clients and suppliers, it becomes even more clear that innovation processes cannot simply move from the firm towards the market. In fact the process has stimuli other than the final demand, found in the relationships with suppliers and distributors and in other horizontal relations (Rullani 1989). The process moves in a circular path, rather than a linear one.

Lastly, university-local environment relations have historically been the least interesting of all regarding innovation, except that these tend to grow where university-firm relationships have also been fruitful. The spin-off processes that move from the university to the free market and transform the invention into innovation are indeed stimulated by the relations of the university within the local context. However here the case is not one of relations with reciprocal utility or relevance, but instead seems to take shape in a unidirectional form, where the university maintains an ample margin of control.

Even if the actors involved in an innovation system interact through the three categories of relationships described here, all actors together still feel the systemic effect that the network of relationships generates around them. It is thus appropriate to now shift the analysis from the interpretive logic of the single relationship to an examination of the dynamics that guide the innovation processes, within the overall systems of relations. In effect we now shift the discourse from "bipartite" reasoning to network reasoning. One of our particular objectives is to understand the discretional margins that might remain to the firm in deciding if and how to participate in an innovation system.

2.4 The Territorial Variable

It is now widely accepted that innovation is a territorially embedded process, which cannot be fully understood independent of the social and institutional conditions of individual places (Lundvall 1992; Asheim 1999). The "territorially embedded" factors influencing the innovation process have thus become the focus for a number of theoretical perspectives (Crevosier 2014): from that of the innovative milieu (Camagni 1995) and industrial district (Becattini 1987) to those of the learning region (Morgan 1997) and innovation system (Cooke et al. 1997; Cooke 1998). These approaches are often interrelated, and as a whole they offer powerful insights that aid our understanding of the mechanisms and necessary conditions for innovation processes (Moulaert and Sekia 2003).

Some of the most relevant findings deriving from these approaches concern the significance of territorial proximity, local synergies and interaction (Camagni 1995, p. 317), and the importance of local inter-organisation networks, financial and legal institutions, technical agencies, research infrastructures, education and training systems, governance structures and innovation policies (Iammarino 2005, p. 499) in shaping innovation.

The "territorial" approach provides a flexible theoretical tool for the identification of a series of external conditions under which learning and innovation occur (Cooke et al. 1997, p. 485), and that can then be applied to analyses across innovation systems and as a basis for actual innovation strategies. These external conditions render some courses of innovation action easier than others (Morgan 2004) or serve as "social filters", meaning unique combinations "of innovative and conservative elements that favour or deter the development of successful regional innovation systems" (Rodríguez-Pose 1999, p. 82) in specific places.

Territories draw on their internal capacities to produce innovation, through direct inputs in the research process or through the creation of innovation-conducive systems in the local environment. At the micro-level of a territory, innovative units (company R&D departments, universities, research centres), as well as local institutions and individuals, interact with each other and with their external environment through the networks of actors described previously. Such interactions stimulate the transmission of knowledge in the form of knowledge spillovers (Jaffe 1986; Acs et al. 1992) that are reaped by the same local actors.

Territories also have capacities to attract and assimilate innovation that has been produced elsewhere. Knowledge spillovers can originate locally but they can also be generated from outside the local or regional borders, as "there is no reason that knowledge should stop spilling over just because of borders, such as a city limit, state line or national boundary" (Breschi and Lissoni 2001; Audretsch and Feldman 2004, p. 6).

The potential for knowledge spillovers beyond regional or national borders depends on the nature of the knowledge concerned. According to Leamer and Storper (2001: 650) codified information can be transferred inexpensively because its underlying symbol systems are readily disseminated through the information infrastructure. It can also be moved over long distance and does not suffer from heavy distance decay in the process.

However, not all information is completely codifiable, and certain features can make codification too costly or even impossible. In these cases, the firm's acquisition of the codification "symbol system" or the physical infrastructure related to an innovation will not be sufficient to achieve successful transmission of the knowledge (Storper and Venables 2004: 354). Instead there is a need to disseminate tacit knowledge through intrinsically "spatial" communications technologies, where face-to-face interaction is at times essential. Face-to-face contacts, as discussed in Storper and Venables (2004) and Charlot and Duranton (2006), not only serve as a means of communication but also fulfil other functions that make communication possible and effective. Face-to-face contacts thus ease the innovation process.

However, in contrast with codifiable information, the transmission processes for tacit knowledge are costly and suffer from strong distance decay effects. Face-to-face contacts are maximised within relatively small territories, because of a combination of proximity and the presence of common socio-institutional infrastructures and networks. The potential to reap knowledge spillovers will thus be maximised within the region. Some of this knowledge will nevertheless spill over beyond the borders of the region or locality and flow into neighbouring areas, due to different forms of inter-regional contacts. Flows of inter-regional knowledge are thus important agents of innovation, but their influence will tend to wane with distance and the accompanying decay of effectiveness in face-to-face and other forms of interaction (Anselin et al. 1997; Adams and Jaffe 2002; Adams 2002).

2.5 Local Innovation Systems in Emerging Nations

During recent decades, newly industrializing economies such as Korea, Taiwan and Singapore have experienced intensive technological learning and made much progress in closing the gap with developed countries (Kim and Nelson 2000; Lee and Von Tunzelmann 2004). This transformation has promoted the growth of a body of empirical literature focused on explaining how developing countries, including the newly industrializing economies (NIEs), have gained ground on the advanced countries, particularly by assimilating and adapting the mature technology of the advanced countries (Vernon 1966; Utterback and Abernathy 1975; Kim 1980; OECD 1992; Dahlman et al. 1985).

A large part of the studies address the process of technological development in emerging nations by adopting a micro-level perspective that compares the innovation process of firms from developed countries with the process adopted by latecomer firms.[11] Beginning from this approach, Kim (1980) argues that the innovation process in emerging nations is fundamentally different from that of developed countries, and he proposes a three stage model to explain innovation processes in emerging nation firms. During the early phase (duplicative imitation) firms acquire mature foreign technologies from advanced countries, including packaged assembly processes that only require some limited local production

[11] Latecomer firms are companies in developing countries that "are able to exploit their late arrival to tap into advanced technologies, rather than having to replicate the entire previous technological trajectory. They can accelerate their uptake and learning efforts, utilizing various forms of collaborative processes and state agencies to assist with the process, bypassing some of the organizational inertia that holds back their more established competitors" (Mathews 2002: 470).

engineering. In the second phase (creative imitation), process development and product design technologies are acquired. In the third phase (innovation), R&D is applied to produce new product lines. According to this model, true innovation occurs only in the final phase of the process, which begins with the simple acquisition and imitation of innovations realised by others.

This scheme seems substantially similar to that of a three-stage model proposed by Lall (1992), who observed the process of innovation development in Indian multinationals. The first stage is based on a learning process that the author defined as "learning within a given technology", based on the imitation of an existing technology acquired from foreign multinationals by means of knowledge gathered through reverse engineering. During a second stage, called "learning by embodied technology", the Indian firms began to modify the products of the first stage through small adjustments that made them better suited to the demands of local consumers. In stage three, the Indian firms took advantage of the knowledge acquired in the preceding stages to develop new capacities to achieve incremental innovations, which they could also export to other developing nations, thus obtaining a lasting competitive advantage.

Hobday (1995, 2005), studying the process of technological development in NIE firms, showed that in such firms process innovations preceded product innovations. The author proposed a three stage model of overall development. The first stage is "original equipment manufacturing" (OEM), in which firms of latecomer nations exploit their low labour costs and assemble standardised products on order and according to instruction from other firms, generally foreign multinationals, who market the products in the rest of the world under their own brand. In the second "original design manufacturing" (ODM) stage, latecomer firms develop autonomous production capacity but the products are still designed and marketed under the brands of multinational principals.[12] In the final "own brand manufacturing" (OBM) phase, the latecomer, thanks in part to continuous investment in R&D, acquires independent capacities for designing and producing new products that they market under their own brands.

Further authors have described how moving from the OEM to OBM phase does not always occur in the same manner. In some cases the learning processes are so rapid that the local firm skips the ODM phase and immediately arrives at creating products under its own brands (Lim and Lee 2001; Lee et al. 2005).

Applying a resource-based view, some authors (Mathews and Cho 2007; Mathews 2002; Huang et al. 2010) have underlined that the core competencies of NIE firms do not derive from possessing specific resources but rather from the capacity to identify innovative firms in developed nations, and then cultivate links with these for acquisition of missing resources and competencies.

---

12 An excellent example is the USB flash drive, largely produced by Chinese firms like Huawei and ZTE, and then branded by local operators under ODM contracts.

Many subsequent authors have based their work on the approaches just outlined, using them to explain technological development of firms from particular geographic contexts, but without adding significantly to the actual theories. These studies have benefited the understanding of the specificity of processes in developing-nations firms, but offer only a partial understanding of the overall phenomena of innovation development in emerging nations. This is because just as in developed nations, the development of innovation in single firms is closely tied to the system of relations in which they are inserted, and is not the result of individual choices that can be made by independent actors, but rather of collaboration among different actors, such as firms, institutions and research agencies, found in the territorial context (Bell and Albu 1999).

Some authors have attempted to explain the innovation development process in emerging-nation firms through analysis of the relations between the technology donor, which is usually a foreign multinational, and the developing-nation company. This category includes studies on the "global value chain" (GVC) (Gereffi et al. 2005) and the "global production network" (GPN) (Ernst 2002), which show that learning in developing-nations firms springs principally from their insertion in the global production network of foreign multinationals, which often externalise some activities of their value-added chain to developing-nations firms in order to reduce production costs. GVC literature stresses the role played by leaders in the chain in terms of transferring knowledge to their suppliers (Giuliani 2005; Pietrobelli and Saliola 2008). For small firms in least-developed countries, participation in value chains is a crucial means of obtaining information on the type and quality of products and the technologies required by global markets, and of gaining access to those markets. However this information needs to be combined with local technological capabilities, which requires substantial technological and learning efforts (Morrison et al. 2008). GVC analysis is limited because of the lack of attention to the institutional context within which local firms are embedded. This limitation is highlighted in the literature on global production networks, which deals with how actors in various networks are embedded, including the geographical dimension from national to local scale (Ernst 2002; Hess and Yeung 2006).

But even the GVC and GPN perspectives offer a limited vision of the innovation development process in emerging-nation firms, since they neglect to illustrate the effects of relations with other types of actors belong to the innovation system, such as government institutions or research centres (Pietrobelli and Rabellotti 2011). While firms are the primary actors in the generation of technological objects, their activities are supported by the accumulation of knowledge and skills in a complex milieu of other research and training institutions. Such systemic factors also apply in developing countries, where technological effort is embedded in the specific economic, policy and institutional context of each country. Policy on technology must necessarily encompass this wider context. The component institutions of the system are of course the private firms, working individually or in collaboration, but also universities and other educational bodies, professional societies and government laboratories and research institutions, private consultancies and industrial research associations (Lall and Pietrobelli 2005).

This aspect has been highlighted by scholars examining the characteristics and specificities of national and regional innovation systems of some developing nations (Intarakumnerd et al. 2002; Patibandla and Peterson 2002; Finegold et al. 2004; Lee and Saxenian 2008). Their contributions are primarily descriptive in character, not offering alternative theoretical models, but have the benefit of showing certain peculiarities of developing nation innovation systems that clearly distinguish them from the models proposed for developed nations (Park and Markusen 1995; Yeung and Lin 2003).

First of all, foreign firms and their direct investments play a much more prominent role in the innovation systems of developing countries than they do in developed countries (Diez and Berger 2005). In a way, the innovation systems of developing countries are significantly globalised: the majority of new knowledge and technologies is acquired from extra-regional sources via direct investment or technology licensing (Ernst 2002). Further, innovation systems in emerging nations generally seem to appear as forms of "learning systems" rather than "innovation systems", since their fundamental aim consists of supporting processes for the learning and acquisition of pre-existing technological knowledge rather than stimulating new production (Mathews 2001; Viotti 2002; Lall and Pietrobelli 2005).

Also, developing-country innovation systems are characterised by the fragmentation of actors and their linkages (Intarakumnerd et al. 2002). Some actors are still missing or incapable of contributing to innovation activities, and embeddedness and trust among these actors are still insufficient to support the evolution of strong linkages. Finally, the State plays a significant role in building successful innovation systems in developing countries (Yusuf and Stiglitz 2001). Basic investments in the innovation infrastructure are necessary, and have to be made by the public sector. The investments include human capital development, academic capacity building, and the promotion of industrial innovation, local content, and technology transfer. The interventionist governments of the so-called developmental States tend to dominate what are in effect centralised innovation systems. At the same time, these governments face the challenge of also allowing competitive forces to take effect.

The role of universities as alternative sources of knowledge appears to be marginalised in developing contexts. While research into the industrialised economies strongly emphasises phenomena such as science-based industries, university spin-off firms and entrepreneurial universities, in the developing countries it seems that the role of universities is less central to national economic upgrading and technological progress. Universities contribute solely to education (Liefner and Schiller 2008; Altbach and Salmi 2011).

These specificities generate much scepticism among scholars about applying the LIS concept to developing countries. The main concerns regard the very concept of innovation, given that developing countries are not able to realise "true" innovation. Some critics also claim that the concept of localization is no longer particularly relevant in a globalised world. However Lundvall et al. (2009) affirm that the LIS concept can be applied to developing countries if we substitute the traditional STI model (Science-technology-innovation) with the DUI model

(Doing-using-interacting), which represents innovation as the result of interactive processes not necessarily based on science and technology. In this sense it is then possible to apply the concepts of NIS, RIS and LIS to developing countries.

Another area of doubt concerns the supposed systemic aspects of the LIS, given that in developing countries the relationships among the different actors of the innovation system are seen as weak or absent (Chaminade and Vang 2008). Chaminade and Vang (2008) thus affirm that the concept of LISs is applicable to developing countries only if they are conceived in an evolutionary perspective that distinguishes between emerging and mature LISs. In the emergent LIS the market transactions dominate the interactions between the building blocks of the system (firms, universities and other knowledge providers and users). Additionally, emergent RISs frequently show weak inter-sectoral links, the absence of university-industry interface structures, and universities that specialise mainly in the supply of manpower (Galli and Teubal 1997). In emergent LISs, firms and other building blocks of the system are still engaged in accumulating the competences and capabilities necessary for different forms of interactive learning. The emergent RIS might gradually evolve into a mature RIS. In the mature RIS, interactions between the building blocks take place through market and non-market mechanisms such as information links, personal interactions, and other kinds of formal and informal networks. Firms and other organizations in the system have developed their absorptive capacity and are engaged in continuous interactive learning with other firms, users, universities and other organizations in the system. It follows that the university–industry linkages are strong and involve various forms of knowledge transfer.

Arocena and Sutz (2000) also call for an ex-ante perspective in the analysis of innovation systems in developing nations, which could identify potential initiatives and trajectories for construction of such systems in areas where they do not exist, rather than the typical ex-post approach of those that focus on developed nations, and which concentrate analysis on the functional logic of pre-existing systems. Consequently the specificity of LIS in developing countries need not be considered as a failure to achieve the traditional model of innovation systems, and should instead stimulate a reinterpretation of the model in dynamic mode (Lundvall 2007).

To understand the dynamics of innovation development in emerging nations, it would be useful to outline a sort of evolutionary route in which the development of a local system of innovation, as it is structured in developed nations, would in fact be the arrival point.

To do this it seems useful to integrate the contributions on developing nations that analyse the processes of innovation for single firms with those on global value chain and local innovation systems, which investigate the relational dimension of innovative process. This integration should concern both the binary relation between developing-nation firms and foreign multinationals, and the broader view referring to the entire network of actors in which the company is inserted. Given the entirety of the theoretical and empirical literature, we can reasonably hypothesise that the traditional configuration of innovation systems, with three typologies of actors (firms, research centres and local institutions), which interact to facilitate the processes of innovation development, only appears in developing nations in the last phase of the innovation process. Further, during the initial phases of acquisition and assimilation, we can hypothesise the existence of other types of systemic structures and interactions of actors. Also, in accord with the global-value chain and global-production network perspectives, we can consider that in the phases that precede innovation, there is an important role for multinational corporations. Such companies are in fact more or less aware of serving as the principle source for developing nation firms to acquire the knowledge necessary to launch their own innovative process.

2.5.1 A Theoretical Model for the Emerging Nation LISs

Integrating the various analytical perspectives summarised in the previous paragraph, we can imagine a sort of ideal path for the constitution of innovation systems, depending on the actors involved and the development level of the local entrepreneurial system.

In the case that the development of the entrepreneurial system is at very low level and demonstrates a lack of actors that can participate in creation of an LIS, the only means to engage such creation mechanism lies in strong government intervention. This implies appropriate policy, aimed above all at attracting the actors lacking in the territory and also favouring their geographic concentration. Government in this case assumes the role not as the sole driver of the system, but as the key actor in guiding the process of LIS creation, in a top-down logic.

In a first stage the government actor implements policies to support the in situ construction of the system of missing actors. Policies to attract foreign actors in particular would only be enacted to guarantee sufficient concentration of the appropriate actors in the given region. Such policies can be directed at both foreign companies and foreign research institutions, although it is usually preferable to target the companies. Attracting companies is easier than attracting research centres and provides immediate results in the creation of local systems.

Once foreign actors are in place, government must also provide specific policies for the generation of mechanisms of knowledge spillover between the newly inserted actors and the local ones. In fact the existence of foreign actors does not engage mechanisms for LIS creation if they do not transfer their knowledge to the local actors. Indeed in many cases government policies have failed, precisely because they ended at providing for the attraction of foreign actors, and did not follow up with appropriate interventions such that these would engage virtuous growth mechanisms for the benefit of the other local actors.

In the case that the local entrepreneurial systems are more developed and a large company already exists in situ, it could be that this firm, pushed by the necessity of broadening its competitive horizons, would engage virtuous mechanism for the creation of an LIS. When large companies localise in a specific territory they create a range of strong and weak links with other local actors, which can contribute to their growth.

The large firms can interact with the other local firms, not only through instituting supply relationships, but also by establishing virtuous mechanisms for knowledge transfer through appropriate cooperative relationships. Large firms can also interact with research centres, through the financing of research or training programmes, and in establishing cooperation mechanism of various types. The large firms can even create other actors in the system through budding processes and stimulation of spin-off companies, and through creation of true corporate research centres and corporate universities.

However the process through which large companies could originate an LIS is not automatic, as might falsely be understood from reading certain literature. Instead it depends on the actual will of the firm, as well as its own capacities for innovation. If the large firm reasons from a purely opportunistic view, refusing to cooperate with local actors, or if the firm is fixed in traditional mechanisms and not sufficiently innovative and proactive, its presence will not guarantee formation of an LIS. Instead, large firms could induce closure of the system on itself, in fact limiting the growth of the other actors in the context (Ferretti and Parmentola 2010).

Thus the creation of the LIS must occur through the cultural growth of the large firm, which aims at the reduction of opportunistic behaviours and continuous stimulation of innovation. This means that the large firm must transfer a concept of cooperation to the local actor: the concept that the growth of all the firms occurs through innovation, and that innovation cannot be realised through an isolated approach, but only through cooperating with other actors.

In other cases the motor for creation of the LIS can be a university, which broadens its mission from that of a research and training institution to that of an actor that guides entrepreneurial development in an area. The enactment of the so-called "third mission" of the university can engage processes of entrepreneurial growth through the creation of academic spin-offs. Spinning off their innovation means that the university acts directly towards the implementation of local entrepreneurial systems. Concession of patents under license or direct commercialisation of research results can also provide a push towards revitalising local firms other than an opportunity to undertake activity with high innovation content. The university can also contribute to the birth of the system through individual cooperation agreements, guiding the firms in the difficult phases of research.

Thus the presence of a university with strong entrepreneurial propensities in an emerging-nation region can contribute to birth of an LIS. However contextual action is generally necessary to ensure that the university structures and personnel decide to undertake the "third mission". In fact in many cases there is resistance from universities and academics against the idea of undertaking entrepreneurial activity, stemming in part from the false conviction that this could reduce space and time for activities of pure research. This conception is not supported by empirical evidence, which shows that the best universities in terms of research results are often those with greater entrepreneurial propensities. The various actors must conceive of and transmit the idea that the "entrepreneurial university" can create benefits for the territory, for firms operating there, and for the university itself, to find new research opportunities and sources of private funding for its own activities.

In summary, the creation of an LIS, in the manner defined in the earlier sections of this chapter, in itself represents a "point of arrival". The structure of the local entrepreneurial system may present a situation where one or more actors are missing, and hence require engagement of a prior creation process that in time can lead to the creation of an LIS. In this sense the LIS concept is applicable in emerging nations only if the key actors share the conception of the dynamic nature of the LIS, as the final point in a process of creation.

In this regard we can imagine the existence of three different configurations of LISs in emerging nations, representing diverse intermediate stages, all in the process of convergence towards the configuration of an LIS as traditionally understood.

The emergence of the different configurations depends on the degree of development of the local entrepreneurial system and foresees the existence of a specific actor capable of engaging the creation process. In this sense, according to the actor involved, we distinguish the three configurations of LISs:

- the government-driven LIS;
- the firm-driven LIS;
- the university-driven LIS.

The government-driven LIS develops in situations where the local entrepreneurial system has a development level such that there is no other actor capable of intervening in the process of system creation. Firm-driven and university-driven LISs take hold in situations where the entrepreneurial system is more advanced.

All three configurations will succeed over time if the guiding actor is capable of engaging positive mechanisms for the involvement of other actors in the creation and construction of a true LIS (see Fig. 2.2).

Subsequent chapters will illustrate these different configurations of an LIS, in part through actual cases that describe their appearance in emerging nations.

Fig. 2.2 Classification of emerging nations LISs

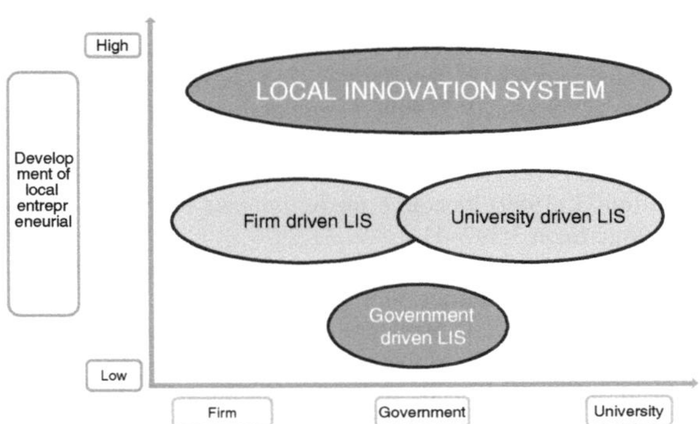

References

Acs Z, Audretsch D, Feldman M (1992) Real effects of academic research: a comment. Am Econ Rev 82:363–367
Adams JD (2002) Comparative localization of academic and industrial spillovers. J Econ Geogr 2(3):253–278
Adams JD, Jaffe AB (2002) Bounding the effects of R&D: an investigation using matched firm and establishment data. Rand J Econ 27:700–721
Altbach PG, Salmi J (eds) (2011) The road to academic excellence: the making of world-class research universities. World Bank, Washington, DC
Anselin L, Varga A, Acs Z (1997) Local geographic spillovers between university research and high technology innovations. J Urban Econ 42:422–448
Antonelli C (1999) La nuova economia della conoscenza e dell'attività innovativa. In: Antonelli C (ed) Conoscenza tecnologica: nuovi paradigmi dell'innovazione e specificità italiana. Fondazione G. Agnelli, Turin
Arocena R, Sutz J (2000) Looking at national systems of innovation from the South. Ind Innov 7(1):55–75
Asheim B (1996) Industrial districts as 'learning regions': a condition for prosperity? Eur Plan Stud 4(4):379–400
Asheim B (1998) Learning regions as development coalitions: partnership as governance in European workfare states? Presented at the second European urban and regional studies conference on 'culture, place and space in contemporary Europe', University of Durham, UK, 17–20 Sept 1998
Asheim B (1999) Interactive learning and localised knowledge in globalising learning economies. GeoJournal 49(4):345–352
Asheim B (2000) Industrial districts. In: Clark G, Feldman M, Gertler M (eds) The Oxford handbook of economic geography. Oxford University Press, Oxford, pp 413–431
Asheim B, Coenen L (2005) Knowledge bases and regional innovation systems: comparing Nordic clusters. Res Policy 34:1173–1190
Asheim BT, Gertler MS (2004) The geography of innovation: regional innovation systems. In: Fagerberg J, Mowery D, Nelson R (eds) The Oxford handbook of innovation. Oxford University Press, Oxford, pp 291–317
Asheim B, Isaksen A (1997) Location, agglomeration and innovation: towards regional innovation systems in Norway? Eur Plan Stud 5(3):299–330
Asheim B, Isaksen A (2002) Regional innovation systems: the integration of local 'sticky' and global 'ubiquitous' knowledge. J Technol Transf 27:77–86
Asheim B, Coenen L, Svensson-Henning M (2003) Nordic SMEs and regional innovation systems. Final report. Nordic Industrial Fund, Oslo. http://www.nordicinnovation.net
Asheim B, Smith HL, Oughton C (2011) Regional innovation systems: theory, empirics and policy. Reg Stud 45(7):875–891. doi:10.1080/00343404.2011.596701
Audretsch DB, Feldman MP (2004) Knowledge spillovers and the geography of innovation. Handbook Reg Urban Econ 4:2713–2739
Baptista R, Swann P (1999) A comparison of clustering dynamics in US and UK computer industries. J Evol Econ 9:373–399
Becattini G (1987) Mercato e forze locali: il distretto industriale. Il Mulino, Bologna
Becattini G (1989) Piccole e medie imprese e distretti industriali nel recente sviluppo italiano. Note Econ 3:397–411
Bell M, Albu M (1999) Knowledge systems and technological dynamism in industrial clusters in developing countries. World Dev 27(9):1715–1734
Braczyk HJ, Cooke P, Heidenreich M (1998) Regional innovation systems: the role of governance in a globalized world. Pinter, London
Breschi S, Lissoni F (2001) Knowledge spillovers and local innovation systems: a critical survey. Ind Corp Change 10(4):975–1005

Camagni R (1992) Economia urbana: principi e modelli teorici. La Nuova Italia Scientifica, Milan
Camagni R (1995) The concept of innovative milieu and its relevance for public policies in European lagging regions. Reg Sci 74(4):317–340
Chaminade C, Vang J (2008) Globalisation of knowledge production and regional innovation policy: supporting specialized hubs in the Bangalore software industry. Res Policy 37:1684–1696
Charlot S, Duranton G (2006) Cities and workplace communication: some quantitative French evidence. Urban Stud 43(8):1365–1394
Clark G, Feldman M, Gertler M (eds) (2000) The handbook of economic geography. Oxford University Press, Oxford
Coe NM, Bunnell TG (2001) Spaces and scales of innovation. Prog Hum Geogr 25(4):569–589
Cooke P (1998) Introduction: origins of the concept. In: Braczyk HJ et al (eds) Regional innovation systems. UCL Press, London, pp 2–25
Cooke P (2001) Regional innovation systems, clusters, and the knowledge economy. Ind Corp Change 10(4):945–974
Cooke P, Uranga MG, Etxebarria G (1997) Regional innovation systems: institutional and organisational dimensions. Res Policy 26:475–491
Crevoisier O (2004) The innovative milieus approach: toward a territorialised understanding of the economy. Econ Geogr 80(4):367–369
Crevoisier O (2014) Beyond territorial innovation models: the pertinence of the territorial approach. Reg Stud 48(3):551–561
Dahlman C, Westphal LE, Kim L (1985) Reflections on acquisition of technological capability. In: Rosenberg N, Frischtak C (eds) International technology transfer: concepts, Measures and Comparisons. Pagamon, New York
de la Mothe J, Paquet G (eds) (1998) Local and regional systems of innovation. Kluwer, Dordrecht
Diez JR, Berger M (2005) The role of multinational corporations in metropolitan innovation systems: empirical evidence from Europe and Southeast Asia. Environ Plan A 37(10):1813
Doloreux D (2002) What we should know about regional systems of innovation?. Technol Soc: Int J. 24:243–263
Doloreux D, Parto S (2004) Regional innovation systems: a critical synthesis. INTECH Discussion Paper Series, 2004–17. United Nations University, Institute for New Technologies, Maastricht
Dopfer K, Foster J, Potts J (2004) Micro–meso–macro. J Evol Econ 14:263–279
Edquist C (1997) Systems of innovation: technologies, institutions and organizations. Pinter, London
Ernst D (2002) Global production networks and the changing geography of innovation systems: implications for developing countries. Econ Innov New Technol 11(6):497–523
Etzkowitz H, Leydesdorff L (1997) Universities in the global economy: a triple helix of government-industry and government relations. Croom Helm, London
Etzkowitz H, Leydesdorff LA (2000) The dynamics of innovation: from national systems and "mode 2" to a triple helix of university–industry–government relations. Res Policy 29:109–123
Fagerberg J (2004) Innovation: a guide to the literature. In: Fagerberg J, Mowery D, Nelson R (eds) The Oxford handbook of innovation. Oxford University Press, Oxford, pp 291–317
Feldman M (1994) The geography of innovation. Kluwer, Dordrecht
Feldman M (1999) The new economics of innovation, spillovers and agglomeration: a review of empirical studies. Econ Innov New Technol 8:5–25
Feldman M (2000) Location and innovation: the new economic geography of innovation, spillovers, and agglomeration. In: Clark GL, Feldman MP, Gertler MS (eds) The Handbook of economic geography. Oxford University Press, Oxford, pp 373–394
Feldman M (2001) The entrepreneurial event revisited: firm formation in a regional context. Ind Corp Change 10(4):861–891
Ferretti M, Parmentola A (2007) The creation of regional innovation systems in emerging countries: the case of Dubai. Int J Globalisation Small Bus 2:137–165

Ferretti M, Parmentola A (2010) Lo sviluppo dei sistemi locali di innovazione nei paesi emergenti. In: Izzo F (ed) Reti per l'Innovazione. McGraw-Hill, New York City
Finegold D, Wong PK, Cheah TC (2004) Adapting a foreign direct investment strategy to the knowledge economy: the case of Singapore's emerging biotechnology cluster. Eur Plan Stud 12(7):921–941
Freeman C (1987) Technology and economic performance: lessons from Japan. Pinter, London
Gereffi G, Humphrey J, Sturgeon T (2005) The governance of global value chains. Rev Int Polit Econ 12(1):78–104
Gertler M (2003) Tacit knowledge and the economic geography of context: of the undefinable tacitness of being (there). J Econ Geogr 3:75–99
Giacometti M (1987) Strumenti di stimolo alla nascita delle imprese ad alta tecnologia: il parco scientifico di Cambridge. L'Industria (2)
Giuliani E (2005) Cluster absorptive capacity why do some clusters forge ahead and others lag behind? Eur Urban Reg Stud 12(3):269–288
Goddard J (2011) Connecting universities to regional growth: a practical guide. European Union, DG Regional Policy, Brussels
Goddard J, Puukka J, Duke C, Dubarle P, Benneworth P (2007) Higher education and regions: globally competitive, locally engaged. OECD, Paris
Granovetter M (1985) Economic action and social structure: the problem of embeddedness. Am J Sociol 91:481–510
Hess M, Yeung H (2006) Whither global production networks in economic geography? Environ Plann A 38:1193–1204
Hobday M (1995) East Asian latecomer firms: learning the technology of electronics. World Dev 23(7):1171–1193
Hobday M (2005) Firm-level innovation models: perspectives on research in developed and developing countries. Technol Anal Strateg Manag 17(2):121–146
Hommen L, Doloreux D (2003) Is the regional innovation system concept at the end of its life cycle? In: Presented at the conference 'Innovation in Europe: Dynamics, Institutions and Values' Roskilde University, Denmark, 8–9 May 2003
Howells J (1999) Regional systems of innovation. In: Archibugi D, Howells J, Michie J (eds) Innovation policy in a global economy. Cambridge University Press, Cambridge, pp 67–93
Howells J (2002) Tacit knowledge, innovation and economic geography. Urban Stud 39(5–6):871–884
Huang J-H, Chou T-C, Le G-G (2010) Imitative innovation strategies: understanding resource management for competent followers. Manag Decis 48(6):952–975
Iammarino S (2005) An evolutionary integrated view of regional systems, measures and historical perspectives. Eur Plan Stud 13(4):497–519
Intarakumnerd P, Chairatana P, Tangchitpiboon T (2002) National innovation systems in less successful developing countries: the case of Thailand. Res Policy 31:1445–1457
Jaffe AB (1986) Technological opportunity and spillovers of R&D: evidence from firms' patents, profits, and market value. Am Econ Rev 76:984–1001
Kim L (1980) Stages of development of industrial technology in a developing country: a model. Res Policy 9:254–277
Kim L, Nelson RR (2000) Technology, learning and innovation: experiences of newly industrializing economies. Cambridge University Press, Cambridge
Kline L, Rosenberg N (1986) An overview of innovation. In: Landau R, Rosenberg N (eds) The positive sum strategy. National Academy Press, Washington
Lall S (1992) Technological capabilities and industrialization. World Dev 20(2):165–186
Lall S, Pietrobelli C (2005) National technology systems in sub-Saharan Africa. Int J Technol Globalisation 1(3–4):311–342
Leamer E, Storper M (2001) The economic geography of the internet age. J Int Bus Stud 32(4):641–666
Lee CH, Saxenian A (2008) Coevolution and coordination: a systemic analysis of the Taiwanese information technology industry. J Econ Geogr 8:157–180

Lee T, Von Tunzelmann N (2004) A dynamic analytic approach to national innovation systems: the IC industry in Taiwan. Res Policy 34(4):425–440
Lee SH, Peng MW, Wang DY (2005) What determines the scope of the firm over time? A focus on institutional relatedness. Acad Manag Rev 30(3):622–633
Lester RK (2005) Universities, innovation, and the competitiveness of local economies: a summary report from the local innovation systems project—phase I. MIT Industrial Performance Center, Cambridge
Leydesdorff L (2000) The triple helix: an evolutionary model of innovations. Res Policy 29(2):243–255
Liefner I, Schiller D (2008) Academic capabilities in developing countries—a conceptual framework with empirical illustrations from Thailand. Res Policy 37(1):276–293
Lim HJ, Lee BT (2001) U.S. Patent No. 6,284,589. Washington, DC, U.S. Patent and Trademark Office
Lundvall BA (1992) National systems of innovation: towards a theory of innovation and interactive learning. Pinter, London
Lundvall BA (2007) National innovation systems—analytical concept and development tool. Ind Innov 14(1):95–119
Lundvall BA, Johnson B (1994) The learning economy. J Ind Stud 1(2):23–42
Lundvall BA, Joseph KJ, Chaminade C, Vang J (eds) (2009) Handbook of innovation systems and developing countries: building domestic capabilities in a global setting. Edward Elgar, Cheltenham
Lyons D (2000) Embeddedness, milieu, and innovation among high-tech firms. Environ Plann A 32:891–908
Marshall A (1930) Principles of economics, 8th edn. Macmillan, London
Martin R, Simmie J (2008) Path dependence and local innovation systems in city-regions. Innov: Manage Policy Pract 10:183–196
Mathews JA (2001) National systems of economic learning: the case of technology diffusion management in East Asia. Int J Technol Manage 22(5–6):455–479
Mathews JA (2002) Competitive advantages of the latecomer firms: a resource-based account of industrial catch-up strategies. Asia Pac J Manage 19:467–488
Mathews JA, Cho DS (2007) Tiger technology: the creation of a semiconductor industry in East Asia. Cambridge University Press
Morgan K (1997) The learning region: institutions, innovation and regional renewal. Reg Stud 31:491–504
Morgan K (2004) Sustainable regions: governance, innovation and scale. Eur Plan Stud 12(6):871–889
Morrison A, Pietrobelli C, Rabellotti R (2008) Global value chains and technological capabilities: a framework to study industrial innovation in developing countries. Oxf Dev Stud 36(1):39–58
Moulaert F, Sekia F (2003) Territorial innovation models: a critical survey. Reg Stud 37(3):289–302
Nelson R (1993) National systems of innovation: a comparative study. Oxford University Press, New York
Noteboom B (2001) Problems and solutions in knowledge transfer. In: presented at 'conference on the influence of co-operation, networks and institutions on regional innovation systems', MPI, Jena
OECD (1992) Technology and economy: the key relationships. OECD, Paris
Oinas P, Malecki EJ (2002) The evolution of technologies in time and space: from national and regional to spatial innovation systems. Int Reg Sci Rev 25(1):102–131
Oughton C, Whittam G (1997) Competition and cooperation in the small firm sector. Scott J Polit Econ 44(1):1–30
Park SO, Markusen A (1995) Generalizing new industrial districts: a theoretical agenda and an application from a non-Western economy. Environ Plann A 27:81–104
Patibandla M, Petersen B (2002) Role of transnational corporations in the evolution of a high-tech industry: the case of India's software industry. World Dev 30(9):1561–1577

Pietrobelli C, Rabellotti R (2011) Global value chains meet innovation systems: are there learning opportunities for developing countries? World Dev 36(7):1261–1269
Pietrobelli C, Saliola F (2008) Power relationships along the value chain: multinational firms, global buyers and performance of local suppliers. Camb J Econ 32(6):947–962
Porter M (1990) The competitive advantage of nations. Macmillan, London
Porter M (1998) Clusters and the new economics of competition. Harvard Business Review, Watertown, pp 77–90
Rantisi NM (2002) The local innovation system as a source of 'variety': openness and adaptability in New York City's garment district. Reg Stud 36(6):587–602
Rodríguez-Pose A (1999) Innovation prone and innovation averse societies: economic performance in Europe. Growth Change 30(1):75–105
Rothaermel FT, Agung SD, Jian L (2007) University entrepreneurship: a taxonomy of the literature. Ind Corp Change 16(4):691–791
Rowe DNE (1987) Science parks as an instrument for technology transfer. Chem Ind 712–715
Rullani E (1989) La teoria dell'impresa: soggetti, sistemi, evoluzione. In: Rispoli M (ed) L'impresa Industriale. Economia, Tecnologia, Management. Il Mulino, Bologna
Saxenian A (1994) Regional advantage: competition and cooperation in silicon valley and route 128. Harvard University Press, Cambridge
Schumpeter JA (1971) Teoria dello sviluppo economico: ricerca sul profitto, il capitale, il credito, l'interesse e il ciclo economico. Sansoni, Florence
Shearmur R, Doloreux D (2000) Science parks: actors or reactors? Canadian science parks in their urban context. Environ Plann A 32:1065–1082
Storper M, Venables AJ (2004) Buzz: face-to-face contact and the urban economy. J Econ Geogr 4:351–370
Swann G, Prevezer M (1996) A comparison of the dynamics of industrial clustering in computing and biotechnology. Res Policy 25:139–157
Todtling F, Kaufmann A (1999) Innovation systems in regions in Europe: a comparative perspective. Eur Plan Stud 7:699–717
Utterback JM, Abernathy WJ (1975) A dynamic model of process and product innovation. Omega 3(6):640–656
Vernon R (1966) International investment and international trade in the product cycle. Quart J Econ 80:190–207
Viotti E (2002) National learning systems: a new approach on technical change in late industrializing economies and evidences from the cases of Brazil and South Korea. Technol Forecast Soc Chang 69(7):653–680
Yeung HWC, Lin GCS (2003) Theorizing economic geographies of Asia. Econ Geogr 79(2):107–128
Yusuf S, Stiglitz J (2001) Rethinking the East Asian miracle. Oxford University Press, Oxford

Chapter 3
Government-Driven LISs

Abstract Governments often assume a significant role in creating LISs, where their absence stems primarily from the lack of having a local entrepreneurial system. Most governments try to create a local system through enacting policies aimed at attracting foreign direct investment (FDI). Nevertheless the presence of FDIs does not automatically generate a positive effect on local economies and favour the birth of LISs. The creation of LISs depends particularly on the realization of spillovers from FDI knowledge. This chapter examines the situation of government-driven LISs, describing the different approaches that governments can adopt to favour FDI knowledge spillovers and drive development of local innovation systems. We then analyse and compare the specific cases of two government-driven LIS, in Iran and Dubai.

Keywords LIS · Knowledge spillovers · Governments policies · FDI · Iran · Dubai

3.1 Foreign Direct Investments and Knowledge Spillovers in the Development of LISs

Local innovation system can be developed in different ways, depending on the roles assumed by the various actors. In the "bottom-up" process, it is the actors of a specific territory that are themselves able to create the local system, by extending their existing activities to involve other actors in the process. Examples of such cases are the local systems born through the intervention of one or more large local companies, which feed the small companies and guide the creation and circulation of knowledge (Porter et al. 2001). On the other hand there are cases of local innovation systems that originate by the initiatives of a research institution, which commercialises its research results and thus activates spinoff—processes that create entrepreneurial firms. The firms collaborate with the research institution and in the long term can grow significantly (Feldman 2001).

M. Ferretti and A. Parmentola, *The Creation of Local Innovation Systems in Emerging Countries*, SpringerBriefs in Regional Science,
DOI 10.1007/978-3-319-10440-9_3

In both cases, public institutions and the State generally assume an important role in the creation of the LIS, but one which is still secondary. Their function is above all in the definition of policies that act on the characteristics of the local context, and can thus facilitate, or hamper, the processes of generation.

However in the emerging nations, the State often assumes a more active role as the motor for LIS development processes, meaning that the system is formed via a "top-down" logic.

In fact one of the problems of emerging economies is the absence of local actors, in the form of companies and research centres that would have the size and capacity to activate a bottom-up development process. In these cases, a nation that has interests in reorienting its economy towards specialisation in high-knowledge sectors can intervene to create the preconditions for development of LISs.

One of the first steps faced in these situations involves the in situ formation of a set of actors that can guide the process of creating an LIS. The growth of such actors is more complex as the local entrepreneurial system of an area is less developed. In the absence of any firms or research institutions specialised in a given sector, the birth of an LIS is impossible. The State thus has to intervene, guiding the birth process through the substitution or insertion of the missing actors.

The solution adopted by different governments has often been to proceed by in situ attraction of actors who operate outside the regional or national boundaries. As these establish in the territory they can activate the creation process for an LIS. Government, in choosing which actors to attract and how to insert them in the context, as well as in acting on the manner that the new institutions interact with the local actors, in fact orients the entire process of LIS formation.

In this case we can apply the term "government-driven", because the LIS is created at the initiative and as a consequence of policies provided by the public institutions. Although government also has the possibility of targeting policies at the attraction of research centres, they have tended to take actions directed at foreign companies. The aim then, becomes that of favouring the transfer of the technological knowledge from these foreign companies to the local firms.

Technology can diffuse between firms and across regions and countries through various transmission mechanisms (Pietrobelli 1996), including: (i) movement of goods through international trade; (ii) movement of capital through inward and outward foreign direct investment (FDI and OFDI); (iii) movement of people through migration, travel, and foreign education of students and workers; (iv) international research collaboration; (v) diffusion of disembodied knowledge through media and internet; (vi) integration into global value chains to benefit from the foreign technology transferred within the supply chain.

For many years foreign direct investment, as a bundle of technological and managerial knowledge and financial capital, has been identified as a leading vehicle for the transfer of advanced foreign technology to developing countries (Dunning 1994; Lall 2003).

Multinational enterprises (MNEs) are regarded as the world's major drivers of R&D. In most cases they are likely to offer training to their local employees, though this can depend on the enterprise or industry. MNEs are also found to have internal incentives for cross-border technology transfer, in order to share technology between the parent companies and subsidiaries (Markusen 2002). Both theoretical and empirical studies have demonstrated that foreign direct investments generate significant positive effects on domestic companies (UNCTAD 2005a). The presence of foreign MNEs, compared to other investments, can boost domestic company productivity in many ways: by training workers and managers who move or spin off from foreign-owned firms and become available to domestic enterprises (Fosfuri et al. 2001); by demonstrating new technologies, providing technical assistance, transferring patented knowledge, and generating imitation opportunities (Dunning 1993); by creating demand for local inputs, increasing specialisation and efficiency levels in upstream and downstream activities, and by generating other positive externalities for local industries (Rodriguez-Clare 1996; Markusen and Venables 1999), and finally by increasing competitive pressures in favour of static and dynamic efficiencies in domestic firms (Caves 1974; Cantwell 1989).

Some empirical analyses demonstrate that the presence of MNEs generates significant positive effects for the host-country economy, such as reductions in unemployment rates, and rises in financial resources, export rates and gross national product (Singh and Jun 1995). All of these effects seem directly and automatically generated by the presence of MNEs. However such effects may well be temporary, in the sense that they vanish if the foreign firms transfer their activities to other countries.

On the other hand, other theoretical and empirical contributions demonstrate that the presence of MNEs can also generate more permanent effects on host country firms, particularly in the form of knowledge spillovers (Smarzynska 2002). The achievement of knowledge spillovers involves different considerations than are the case for the more "automatic" short-term results of FDIs (Wang and Blomstrom 1992; Rodriguez-Clare 1996; Zhang et al. 2010).

Blomstrom and Kokko (1995) define "knowledge spillover" in the foreign investment context as the voluntary or involuntary transfer of technological and market knowledge from the foreign investors to the local firms. Javorcik (2004) defines firm-level knowledge spillovers as knowledge created by one firm (a multinational enterprise) that is used by a second firm (a host-country firm), and where the host-country firm does not fully compensate the multinational enterprise. In fact in knowledge spillover, the spread of knowledge from foreign MNEs to local, host country competitors is generated by non-market transactions (Meyer 2004).

Blomstrom and Kokko (2001) identify two categories of knowledge spillovers: productivity and market-access spillovers. Productivity spillovers occur when the entry or the presence of an MNE subsidiary generates productivity and efficiency benefits in the host country, such as motivation for product and process innovation, or access to new technology through reverse engineering and labour mobility. On the other hand, market spillovers are benefits that local firms obtain through participation in export of products produced through the local MNE subsidiaries. In implementing such export activity they start "looking abroad", then independently take advantage of the links to international distribution systems and market facilities that have been implemented by foreign firms in the local context.

As is well recognised in the literature, there are three significant "channels" for spillovers of knowledge from multinational enterprises to local firms (Saggi 2006; Liu and Buck 2007).

First, incoming FDIs have a demonstration effect on local R&D activity. With their mere presence in the domestic markets, they bring foreign products and technologies that stimulate and inspire local innovators to develop new products and processes, including through reverse engineering (Cheung and Lin 2004; Hale and Long 2006).

Second, spillovers can be realised through workforce mobility. Hiring and re-hiring practices can permit local firms to obtain the technological know-how of FDI-related firms through "stealing" their skilled workers (Görg and Strobl 2001; Filatotchev et al. 2011).

Finally, spillovers can be achieved through vertical and horizontal linkages between foreign and local companies. Examples of these would be the sharing of know-how, the provision of staff training by the MNE, joint research projects and other activities in common (Blomstrom and Sjoholm 1998; Sjoerd et al. 2008).

The generation of knowledge spillovers through these three channels depends on two sets of conditions (Wang and Blomstrom 1992; Liu and Wang 2003; Smeets 2008): first the FDI motivations, and second the extent to which local companies respond positively to technology gaps and invest in learning activities.

Some authors have illustrated how FDI motivation determines the potential for generation of local knowledge. In fact in some cases, MNEs can be engaged in technology-seeking FDI, where knowledge spillovers are expected to flow from local firms to the multinational enterprise rather than the other way around (Driffield and Love 2007). Also, if the foreign MNE delocalises only its lower value-added activities to the host market, the knowledge spillover are less intense and less useful for local firm development (Feinberg and Gupta 2004).

Regarding the response of local companies, many authors argue that host-country firms need some minimum amount of absorptive capacity to be able to capture knowledge spillovers (Castellani and Zanfei 2003; Lim 2005; Lai et al. 2006). Such absorptive capacity is created by investment in R&D and human capital. The investments ensure that there is a sufficient basis of knowledge and technology to assimilate and exploit the external knowledge (Cohen and Levinthal 1990).

In short, spillovers stemming from FDIs represent one of the main sources of knowledge for companies in emerging and less-developed countries. Nevertheless it is not easy to capture the benefits of this kind of knowledge, because FDI knowledge spillovers occur only if:

- high tech companies are induced to implement FDIs in the country, and to enter in cooperation with local companies;
- knowledge diffusion and circulation policies are implemented, favouring generation of the "demonstration" effect;
- labour mobility between foreign subsidiaries and local companies is not hampered;
- vertical and horizontal linkages exist between investing companies and local ones;
- local firms have the required level of absorptive capability.

Given an understanding of these conditions, host-nation governments can enact policies that influence them, and thus the generation of FDI knowledge spillovers (Ethier and Markusen 1996).

3.2 The Role of Host-Government Policies

Most governments are aware of the positive effects of investment from foreign multinationals, and so have introduced FDI attraction policies (Buckley and Casson 1985; Nunnenkamp 2004). Many authors have analysed and classified the types of these policies.

Brewer (1993), for example, identifies two types of government policies influencing FDIs. First there are policies that encourage potentially incoming multinationals by promoting market imperfections, such as weaker antitrust provisions, undervalued currency and increased asset protection. Second are policies that can have positive effects on MNEs without necessarily generating market imperfections, such as privatisation of trade, adoption of currency convertibility, and reduction of favouritism towards domestic firms.

According to Hanson (2001), the most common FDI attraction policy is the introduction of partial or complete exemption from corporate taxes and import duties for foreign investors, matched by requirements that the multinationals establish production facilities in the host country, at times in designated areas or specified lines of activities.

A number of scholars have studied the experience of specific countries in terms of these and other relationships between host government policies and FDIs (Globerman and Shapiro 1999; Lim 2005; Meyer and Vo Nguyen 2005).

However, as noted in the previous section, the presence of MNEs does not automatically generate knowledge spillovers for local firms (Crespo and Fontoura 2007). In fact the possibility of positive effects depend on the characteristics of the investing companies, on the absorptive capacities of the host-country companies, and finally on the proper functioning of the three spillover "channels".

In this regard, we can distinguish between generic FDI-attraction policies and other types of policy aimed not only at attracting MNEs, but also at increasing the intensity of knowledge spillovers.

Although the generation of knowledge spillovers is one of the few means available for emerging nations to attain real economic development, few scholars have actually analysed the effects of host-government policies on the generation of FDI knowledge spillovers.

Blomstrom et al. (1998) identify three areas of government intervention that can influence the spillover effect: policy on inward FDIs, and trade and technology policies. Policy on inward FDIs defines the levels of constraints imposed on foreign firms, which negatively influence the generation of knowledge spillovers. The characteristics of trade policy typically descend from the host country's import regime. A regime that is "open" to imports can stimulate capture of spillovers by facilitating the import of the technology embodied in capital goods and intermediate inputs, and by encouraging competition from domestic industries. Finally, policy on technology also has important potentials for impact on FDI spillovers. For example, government policies that encourage investments in R&D activity within the host economy can enhance the technical capacity of local firms, and as a consequence their capability to exploit appropriable foreign technology.

In a different perspective, Te Velde (2001, 2002) distinguishes government actions in three categories. These are policies: (i) affecting potential foreign investors; (ii) affecting established foreign investors, and (iii) affecting the response of domestic firms. The coexistence of the three different types of policy supports the generation of positive spillovers. In particular, the first type of policies, generating location advantages, encourages foreign firms to choose a particular country to establish their subsidiaries. They include actions at the industrial level, such as financial and tax incentives, efficient administrative procedures and development of export platforms; and at the macroeconomic level, such as infrastructure development, privatisation policies, and development of national financial markets On the other hand, policies affecting the long-term permanence of FDIs include, at the industrial level, taxation, interaction with research institutions and training of employees; and at the macroeconomic level, labour market conditions, trade and competition policies. Finally, polices affecting the response of domestic firms are aimed at increasing their absorptive capability, strengthening their technological knowledge and human resources skills, promoting direct linkages with multinationals and encouraging labour mobility and supply-side management.

In many emerging nations, government does not limit its interventions to policies that influence economic activity. They also become directly involved through State ownership of companies (Austin 1990). However, in the literature on government FDI policies there is little reporting on the opportunities stemming from direct participation in international business relationships, and little evidence of the impact of such direct involvement on FDI knowledge spillovers. Some authors in fact describe the presence of State-owned enterprises as a disincentive for the entrance of foreign firms (De Castro and Shulze 1995; Meyer and Vo Nguyen 2005; Buckley et al. 2007). On the other hand, other scholars (Narula and Dunning 1998) highlight that, especially for emerging economies, direct government intervention can ensure increase in the knowledge base for both the firms directly linked to foreign partners and for the whole economic system.

The government's assumption of the role as direct partner, in the absence of suitable private companies, may also permit it to better monitor the utilisation of resources and to serve as an honest broker, preventing the misallocation of funds by the commercial partners (Narula and Dunning 1998).

3.2.1 Types of Government Policies for FDIs

Given the theoretical approaches and empirical evidence described in the preceding section, we can understand that different types of government FDI policies exist and

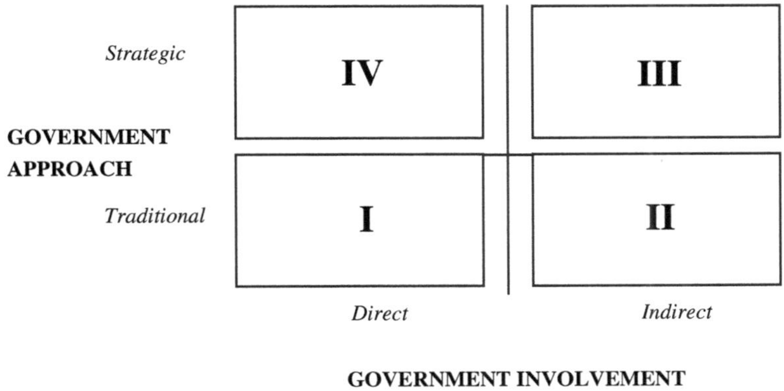

Fig. 3.1 Classification of government foreign direct investment policies

that these involve various levels of State participation, and have differing effects on the generation of knowledge spillovers (Ferretti and Parmentola 2010). To clarify the varying facets of this information, we classify the various government policies according to two broad variables (Fig. 3.1):

- *government involvement*;
- *government approach* to foreign direct investment.

The first variable describes government participation in FDI as:

- *direct*, if the government, through State-owned firms or public institutions, takes part in FDIs as a shareholder in an international joint venture, or more generally as a partner in an international alliance;
- *indirect*, if the government uses indirect instruments, such as contextual legislation, incentives, tax policies, in order to influence FDIs.

The "government approach" variable describes the government use of policy to generate knowledge spillovers, distinguishing two attitudes, as follows.

- *Traditional approach*: The government only adopts policies serving to attract FDI for achievement of short term results, such as reduction of unemployment rates, increase in GNP, supply of financial capital and increase of export level. Government does not attempt to create conditions for generation of knowledge spillovers.
- *Strategic approach*: The government creates the conditions to generate knowledge spillovers, using policies necessary to increase the absorptive capacity of local firms and to motivate the linkages of foreign firms with local suppliers. The approach is called strategic because the government adopts a long term perspective.

From the intersection of the two variables, we obtain a classification of four types of government policies for foreign direct investment (Fig. 3.1), as described below.

Quadrant I: includes all policies characterised by direct government involvement and a traditional government approach to FDI. One example of such policy would be the institution of a joint venture between a State-owned enterprise and a foreign investor, aimed at providing financial resources for a privatisation process or saving local firms from bankruptcy. In this case the foreign partner is only a supplier of financial resources and the potential opportunities of knowledge transfer are ignored. Specific cases of this policy type include the measures initially adopted for the privatisation process in the Russian Federation.

Quadrant II: includes all policies characterised by indirect government involvement and by a traditional government approach to FDI. Examples of this policy are the creation of particular benefits for foreign investors, such as financial and tax incentives, efficient administrative procedures, export platforms, or the definition of favourable prices for productive resources, including labour and raw materials. These policies do not support the generation of new knowledge and in fact could encourage opportunistic behaviour by the foreign investor, in that the company would once again transfer its activities abroad if another country were to offer better cost conditions. These policies have been adopted widely by governments of some emerging nations.

Quadrant III: includes all policies characterised by indirect government involvement and by a strategic government approach to FDI. This quadrant includes government policies intended to increase the absorptive capacity of local firms, such as incentives for human resources training or interaction with research institutions, as well as policies intended to establish long term linkages between foreign investors and local firms, such as laws encouraging labour mobility and supply side management. The goal of these measures is to create knowledge spillovers, thus they are often supported by further policies that encourage the entrance of high-tech firms. Examples of these policies are the measures adopted in recent years by the People's Republic of China, oriented both to increasing the level of technical and managerial knowledge of the local managers and to encouraging joint ventures between local firms and foreign high-tech investors.

Quadrant IV: includes all policies characterised by direct government involvement and by a strategic government approach to FDI. Examples of this category include joint ventures between foreign investors and state-owned enterprises or public institutions, aimed at transferring managerial and technological knowledge from the foreign partners to local firms. Usually in these types of alliances the local partners provide financial resources while the foreign firm transfers its technical and organisational capabilities to the new joint company. This policy type, unlike the others described, promotes the generation of knowledge spillovers not only for the local firms directly linked to the joint venture, but also for the whole economic system.

It is obvious that the different FDI policy types described are not exclusive. In fact a government may first adopt traditional tax policies for the attraction of foreign firms, as well as other policies necessary to then increase the absorptive capacity of the local firms and to motivate the linkages with local suppliers. However for a "short-term, traditional" policy to generate knowledge spillovers, it must feed into the second "strategic" policy.

In summary, only the policies characterised by a strategic government approach to FDI (quadrants III and IV) can directly promote the generation of knowledge spillovers, and thus initiate the creation of local innovation systems.

In the remaining sections of this chapter we provide detailed case studies of Iran and Dubai. These illustrate two different typologies in the implementation of local government policy to favour formation of an LIS. Although the two cases are not directly comparable, they share significant elements:

- the lack of a local entrepreneurial and managerial system that would be capable of initiating and supporting the birth of an LIS;
- the conclusion that the attraction of FDIs would be the most effective solution for LIS creation;
- strong motivation in the local government to create an LIS, stemming from a desire to diversify the local economy, which in turn translates into the adoption of a strategic approach in policies for the attraction of FDIs.

3.3 Direct Government Involvement in Creation of LISs in Iran[1]

The Iranian economy is dominated by the oil industry, in which the operators are State-owned companies. Furthermore, under the Constitution, the State owns and administers large companies in other areas, such as banking, power generation and communications. The private sector includes automobile, textile, metals manufacturing and food-processing companies, as well as thousands of small-scale enterprises such as workshops and farms. A dominant share of Iran's non-petroleum industrial output is controlled by semiprivate charitable organisations called *bonyads*, which have a considerable influence on national economic policy through their close linkages to powerful politicians. Traditional import-export merchants, collectively known as "the bazaar", also have an influential role in economic policy making.

---

[1] The data on the Iranian context have been collected from multiple sources, enabling stronger substantiation of constructs by triangulating the evidence from several fonts. Information was first obtained from a systematic review of trade publications, industrial analyses and reports, business journals, IDRO annual reports, Industrial Management Institute (IMI) reports, websites and reports of foreign multinationals involved in joint ventures. These data were used to describe the overall Iranian FDI policy, illustrate the IDRO goals and sketch an overall picture of the phenomena analysed. A second class of information came from interviews with the senior managers of the IDRO subsidiaries involved in international joint ventures, including a number that had taken part in the UNIDO training program "Institutional Capacity Building for Technology Management at Enterprise Level". The interview results allowed us to describe the international joint ventures. We then interviewed the Deputy for Advanced Industries Development at IDRO, to obtain a systematic picture of IDRO innovation policies and goals. The information from the interviews with the subsidiary managers was compared to that obtained from the Deputy for Advanced Industries Development, to better identify the business model adopted by the Iranian Government for favouring generation of FDI knowledge spillovers.

Recent economic reform programs aim at making Iranian industry more internationally competitive and less dependent on the oil industry. The most important economic reforms include the privatisation of government enterprises, the creation of an equitable income tax system, and the reduction of tariffs that protect local manufacturing from foreign imports (Ghazinoory et al. 2009).

Recently, Iranian policy has also been to gradually open the economy to foreign trade and investment. Since 1996 the United States has imposed the Iran-Libya Sanctions Act, providing for a full trade and financial embargo against Iran. However other countries, including members of the European Union (EU), have continued to trade with Iran (UNCTAD 2005b).

Iran is currently a very attractive country for foreign investors thanks to its unique geographical location, at the cross-roads of the Middle East, Asia and Europe. The nation represents a very substantial market in itself, with a population of 65 million, currently in rapid growth. Iran is an interesting investment focus for its trained and efficient manpower resources, at very competitive cost, and perhaps above all for its varied and plentiful reserves of natural resources, ranging from oil and gas to metallic and non-metallic materials. Moreover, in 2002 the government ratified the Foreign Investment Promotion and Protection Act (FIPPA), which adopts policies aimed at increasing FDIs. These measures include tax incentives, the creation of a number of Special Economic Zones, and improvement in transport and communications infrastructure. The goal of Iranian FDI policies is to attract foreign capital, and above all to draw on the investment to generate technological and market spillovers that drive the creation of LISs. For this purpose the main instrument of Iranian government is the Industrial Development and Renovation Organisation of Iran.

3.3.1 IDRO: The Industrial Development and Renovation Organization of Iran

The Industrial Development and Renovation Organization of Iran (IDRO) was established in 1967 to develop the industrial sector and accelerate the country's industrialisation process, and has since become Iran's largest conglomerate. The agency is fully responsible to the national Ministry of Industry, and in turn controls a large number of private companies, either directly or through its numerous subsidiaries. In the listing of Iran's top 100 companies by size, the first positions are all held by IDRO-controlled firms. Since its foundation, one of IDRO's chief goals has been to change the national economic structure through the development of new and advanced technologies, and thereby favour the globalisation of Iranian industries.

The primary means of obtaining this goal is through implementation of policies aimed at increasing the absorptive capacity of local companies. For example, IDRO controls the Industrial Management Institute (IMI), a subsidiary that serves as an executive education centre. The centre also organises training courses for local managers, which include foreign professors as instructors and faculty. In addition, IDRO allocates 25 % of its annual profit to R&D investment within Iran.

Beside these policies, IDRO has implemented complex mechanisms aimed at acquiring knowledge from foreign companies. These mechanisms arise from the national government's implementation of the Foreign Investment Promotion and Protection Act, which enhances and increases the incoming FDIs to Iran. Given the potential arrival of FDIs, IDRO has implemented its own specific policies that favour the achievement of knowledge spillovers.

The policies include:

- promotion of local and foreign investment in joint ventures with minority holding from IDRO (less than 50 % of shares), with particular emphasis on new, hi-tech and export-oriented industries;
- restructuring of existing industries through participation of reputable foreign companies, in order to transfer new technologies and enhance non-oil exports;
- development of general contracting activities with the participation of Iranian private enterprises and high tech foreign companies;
- offering of consultancy and support services to foreign investors;
- partial privatisation of its existing subsidiaries, with sales of shares to foreign investors.

In order to better illustrate how these policies are implemented we provide detailed analyses of three cases of IDRO joint ventures with international investment.

3.3.1.1 AID-Renault Joint Venture

Iran's two largest auto makers, Iran Khodro and Saipa, are both affiliates of IDRO. In early 2004, they formed a joint company with IDRO called AID Co. The objective was to provide a large domestic car manufacturing company capable of responding to increasing domestic demand. After extensive market analysis, the AID managers concluded that the company could not alone produce vehicle models sufficient to meet the needs of typical Iranian customers. AID then conducted an intensive analysis of a number of foreign auto makers as potential partners. They concluded that the Renault X90 vehicle platform was the most suitable for commercialisation in the local market, and based on this AID and Renault set up a joint company called Renault Pars in May 2004. The aim of the agreement was the joint production and marketing of the low-end "Logan" model, based on the X90 platform, targeted at both Iranian and other emerging markets, with production to begin in 2006. Shares in the joint venture are held 51 % by Renault and 49 % by AID. Iran Khodro and Saipa assemble the X90 vehicle using a mix of components, both imported and produced locally under Renault Pars. In the case of the local production, the agreement requires that most subcomponents must be supplied by local subcontractors. The joint venture is responsible for all production operations, including engineering, quality control, purchasing and logistics, and also coordinates sales policy, marketing and after-sales service. Under the AID-Renault agreement the X90 model is sold under the Renault brand, but through the Iran Khodro and Saipa dealer networks.

This policy has had positive effects. In 2009, Iran Khodro and Saipa held 60 % of the domestic automobile market, in spite of increased competition from off-shore imports. The technological learning generated through the Renault alliances has also recently permitted the Iranian auto makers to design and build completely home-developed automobiles. Even though these vehicles are not as advanced as the foreign cars on the Iranian market, this still represents a technological breakthrough. Saipa was the first to present a completely designed and domestically manufactured car, the Tiba (Gazelle) in December 2008. In April 2009, Iran Khodro announced their own completely domestically-produced automobile, the Rana or Navand, for commercial production beginning in 2010. The Iran Khodro assembly plant near Dakar has also produced several hundred cars specifically for use as taxis in Senegal. In May 2010, Saipa opened a large automotive assembly plant in Kashan, Iran. This plant is the largest auto assembly factory in the Middle East, with a production capacity of 150,000 vehicles per year. At full capacity the plant will add roughly 15 % to Iran's auto production capacity. The plant is entirely Iranian-designed, although 40 % of the machinery was imported, and is currently focused on production of the Tiba car model (Alizadeh and Hakimian 2014).

3.3.1.2 MANA-OTEP Joint Venture

The IDRO Construction Company (MANA) was established in 1974 with the goal of planning, engineering, and constructing large industrial, commercial, administrative, residential and hydroelectric projects. Since its establishment, MANA has completed more than 200 large projects including civil-sector steel mills, machine manufacturing plants, agro-industrial plants and power plants, as well as various residential complexes, State administration and educational facilities. MANA is one of the ten largest construction companies in Iran and is also active in overseas construction projects.

Construction is among the most challenging sectors of the Iranian economy, both because of the substantial investments necessary for new projects and because it is an area demanding continuing innovation. The sector also faces two critical situations. One of these is the continuing increase in national and urban population, necessitating creation of new settlement areas and the development of new civil infrastructure. The other is the repeated instance of high-intensity earthquakes, requiring the re-engineering of existing buildings and the construction of new ones to higher standards of resistance (Aadal et al. 2014). Given the needs to acquire new sectorial technological knowledge, in December 1995 MANA entered into agreement with OTEP SA, a French company, to establish a joint venture company for the production of precast/prestressed concrete components in Iran.

OTEP was not chosen casually, but as the result of a deep analysis of the concrete industry. OTEP was founded in the 1960s, by a team of engineers and technicians with previous experience in the concrete industry, and developed significant and unique technological expertise and know-how in the particular area of precast, prestressed concrete. OTEP's technology and equipment permit cost-effective production of prestressed components for a range of structural and industrial applications. These achieve considerable savings of money and time compared to preceding or "conventional" construction methods.

OTEP and MANA met for the first time in 1995, in Iran. Following this, MANA representatives visited OTEP production facilities in France. MANA concluded that similar factories for precast products could be built in Iran and neighbouring countries, and furthermore that most of the plant equipment could be manufactured in Iran under OTEP instruction and responsibility. In 2004 OTEP accepted a long-term cooperation agreement with MANA, and within several months the two companies established a joint venture called OTEP-IRAN. One of the goals of the joint venture is to supply and install the specific OTEP-designed technology for production of precast, prestressed structural components and light poles in Iran and its neighbouring countries. The agreements specify that such equipment must to the greatest extent possible be manufactured in Iran.

The cooperation with OTEP has permitted the local company to acquire new knowledge in construction methods, particularly manufacturing of prefabricated units, which has historically been one of the most successful solutions in the local construction industry. The success of prefabricated systems in Iran is due to a series of factors, such as the lighter weights compared to conventional construction, capacities for rapid assembly, improved seismic resistance, the economy of the final product, ease in quality control, and the flexibility these systems permit in architectural design (Mirzaei 2012).

3.3.1.3 AMID-Technochem Joint Venture

Advanced Materials Industrial Development Company (AMID Co.) is a subsidiary of IDRO. Its primary goal is the establishment of the advanced materials industry in Iran. Over the years, AMID has invested heavily in the field of advanced materials but still observes a lack of sufficient knowledge in the local context. The company has thus determined to search for foreign partners.

After the deep analysis of a number of a selection of companies active in advanced materials production, AMID decided to implement a joint venture with Hanyoung Technochem Co. Technochem is a Korean firm active in research and development of new refrigerants, as replacements for ozone-depleting substances. Examples are pure and mixed hydrocarbon refrigerants in place of CFCs and HCFCs, transfer fluids for heat pipes, refrigerants for self-chilling cans, and CF3I for fire suppression.

The Persian Refrigerant Development Company (PRD Co.) was formed as a joint venture, held 67 % by AMID and 33 % by Technochem. Technochem provides patent licenses, management training, and project management. PRD produces a new generation of environmentally-sound hydrocarbon-based refrigerants used in applications such as refrigerators, foaming agents and auto parts. The market for this aspect of PRD's production consists of other local manufacturing companies, which apply the new materials in improving the quality and environmental aspects of their own product outputs. Specific areas of development by PRD include pure and mixed hydrocarbon refrigerants for CFCs and HCFCs, CF3I for fire suppression applications, and blowing agents for foam production. PRD has also developed alternative energy products, components for thermoelectric cooling and generating (TEC, TEG) systems, and high energy-efficiency heat pumps.

Summarising from the above three case descriptions, we can identify the common lines of the IDRO policy.

In all cases, an IDRO subsidiary carried out a partner selection process with technological and business due diligence. All of the selected firms either have a strong reputation in specific industries, such as Renault, or possess particularly innovative solutions, such as OTEP and Technochem.

IDRO itself, or one of its specialised sub-holdings, then promotes the start-up of the equity joint venture, providing legal support, financial resources, bank credibility, well trained managers and market access through its wide networks. In exchange it takes a stake in the company and the right to nominate one member of the board of directors. The foreign partners, as part of the start-up agreement, are required to provide technical know-how, technology transfer and industry specific management skills.

This policy achieves the traditional first step of attracting foreign firms, which perceive opportunities for operation in large unexplored markets, with requirements for minimum capital expenditure. However it also favours the next step of generation of knowledge spillovers for both the IDRO subsidiary itself and the surrounding national system. By adopting this strategic model, IDRO forces its international partners to transfer their knowledge to local partner. First of all, the approach enables growth in local enterprises, which acquire new knowledge and technological solutions, becoming more competitive both on local and foreign markets (see the AID-Renault case in particular). Local firms can then also function in bridging processes of knowledge transfer, through spillovers towards firms that serve as suppliers (AID-Renault and the MANA-OTEP cases) or through purchasing and applying the high-tech product as customers of the joint venture (the AMID—Technochem example).

The approach adopted by the Iranian government has certainly achieved improvement in the knowledge and competitiveness of local enterprises. Through attainment of new knowledge, the firms have become competitive in their sectors and are able to outperform the international competition (the AID-Renault case) or achieve new technological solutions that respond to local markets (MANA-OTEP and AMID-Technochem cases). The specificity of the Iranian model is that it is the government itself that collaborates with the foreign companies, through its participation in joint ventures. Obviously it is still not possible to identify the existence of true local innovation systems in the cases presented, but the Iranian policies have established the preconditions for the development of local enterprises at high technological levels, and have thus contributed a useful first element for the future birth of LISs. Theoretically, the fact that the new innovative firms are State-owned should stimulate greater propensity to cooperate with other actors that share a basic orientation towards the collective good, not only with individual companies. This aspect could be a further element in favour of the future development of a local system of innovation.

3.4 Indirect Government Involvement and the Creation of LISs in Dubai[2]

The United Arab Emirates (UAE), formed in 1971, is the only Arab nation with a federal government system. The nation has maintained high rates of development since its formation and the overall economy is now one of the largest and most dynamic of the Gulf region. Among the seven emirates in the federation, Abu Dhabi and Dubai have the largest economies.

Estimates from The Economist (EIU Country Report, April 2011) indicate that in 2009 the Arab Emirates GNP fell 2.7 %, largely due to a 10 % cut in petroleum production, agreed under OPEC, as well as to contraction in the real estate market and the international financial crisis. Per capita GNP is estimated at 33,635 US dollars. Forecasts for 2010 growth varied between 1.5 and 2.5 %, and the estimate for 2011 was 3.5 %.

In spite of recent success in diversification in some of the emirates, particularly Dubai, the federation's economy continues to be highly dependent on petroleum and natural gas. Over the years, Dubai has been the most active of all the emirates in policies for achievement of an economic model that is not oil-dependent. Dubai does produce petroleum but has less deposits than other emirates. This particular factor has served as a stimulus for local government to invest in a number of new sectors (Ewers and Malecki 2010).

[2] For the case of the UAE we have analysed the configuration of three technology systems developed in the Dubai area (Dubai Internet City, DuBiotech, Dubai EnPark), which are intended to evolve towards true local innovation systems. We identified the actors that compose each system, distinguishing the local firms, international firms, universities and research centres. The data concerning the actors present were gathered through documents furnished by the directors of the three science parks. For each actor, we next identified the operational sector and the specific activities carried out in Dubai. This analysis permitted definition of the role assumed by the various subjects, and so the identification of those that carry out the activities most oriented towards knowledge production, thus assuming the role of key actors. Next we reconstructed the relations between the various subjects. This analysis was carried out by combining data obtained from the documentary sources and information from unstructured interviews with the directors of the individual parks, conducted between 2006 and 2010.

In the 1980s the government adopted strong incentives and development policies for the financial services and tourism sectors. In a short time, Dubai became the leading Middle East financial centre, in part by positioning itself as a platform for delocalisation of western multinationals' off-shore financial activities. In regard to the tourism sector, given that the territory held little attraction in tangible cultural-heritage assets, the policy target was to develop a series of super-luxury infrastructure projects for business and leisure clientele.

The true turning point began in the 1990s, when the government decided on a definitive policy of economic diversification, aiming primarily at technology-intensive sectors such as nanotechnology, biotechnology and information-communications technology (ICT), with the ambition of full entry in the new knowledge economy. A further milestone in policy development was the adoption of the Dubai Strategic Plan, signed in 2005 for implementation by 2010. This was in turn followed by a 2010–2015 strategic plan.

The 2010 plan provided a path toward development in three phases:

- In phase one, government objectives were to reinforce traditional economic sectors such as commerce, transport and tourism;
- During phase two there was to be promotion of activities considered as support for the intensive knowledge sector, such as education, telecommunications, financial services;
- In the final phase the objective was to develop sectors of nanotechnology, biotechnology and renewable energy, while reinforcing research and higher education.

In undertaking the third development phase, government faced a serious initial obstacle in the lack of an entrepreneurial system with sufficient competencies to realise highly knowledge-intensive activities. Therefore during recent years, local government has promoted a policy for attracting direct foreign investment. It has created a total of 28 free trade zones, of which 11 are in development and 17 are fully operational, the most important being Jebel Ali Free Zone (Jafza), near the airport 20 km south-east of Dubai City. Firms localised in Jafza enjoy a series of advantages, such as possibilities of forming totally foreign-capitalised firms, full convertibility of currencies, possibilities for full repatriation of profits, and up to 100 % tax exemption. These policies have encouraged attraction of approximately 6,000 foreign firms from over 80 nations.

3.4.1 Policies for the Attraction of High-Tech FDIs: The Dubai Cities

In addition to attracting multinational firms, the development of innovation also requires promotion of appropriate research activity, however the universities and local research centres in Dubai did not seem capable of sufficient effort in this area. Thus the government, through a State holding company called Tecom Investment, has instituted 12 investment zones that are uniquely specialised in innovative activity. The objective is to attract high-tech firms, but also research centres and advanced universities. Of these, the Dubai Knowledge Village and Dubai International Academic City are particularly notable for their role in hosting information centres and foreign academic institutions, acting as incubators for the entire area.

To understand if the approach adopted by Dubai has been effective in order to contribute to creation of true local innovation systems, we examine the evolutionary path followed by the government-created investment zones that focus on carrying out intensive technology activity, and observe whether some of these could potentially become true local innovation systems.

3.4.1.1 Dubai Internet City

Dubai Internet City (DIC) is intended as a strategic platform for localisation of ICT firms. These can use the Arab Emirates location to access numerous international markets, such as nations in the Mid-East, Asia, central and southern Africa. Launched in 2000, DIC now hosts more than 100 firms, including Microsoft, Cisco Systems, IBM, HP, Dell, Siemens, Sun Microsystems, Computer Associates, PeopleSoft and Sony Ericsson. These international giants are flanked by numerous small and medium firms, including local firms, which support activities of the larger firms.

The area offers a series of advantages for localisation, such as very advanced infrastructure, and consulting and other support services. Firms that localise in the area receive a series of advantages, including possibilities to form fully foreign-capital firms, reduced taxation, and easier access to credit and capital markets. Firms that establish a seat in the Internet City can adopt a company structure as the branch of a foreign firm, the branch of a firm that has a headquarters in another area of the UAE, or limited liability company. Whatever the choice, procedures for establishment are extremely simplified: foreign firms can submit certified documents to the UAE embassy in their home nation, which then forwards them to the Arab Emirates to start the procedure. Firms wishing to invest in the Internet City have the particular advantage of using "firsteps@DIC", a programme that offers office rentals in the DIC for limited periods, to explore the market and decide on the issues for permanent localisation in the area.

Firms in the DIC cluster are active in a variety of ICT sectors, such as software development, IT services and development of telecommunications hardware and installations. Large multinationals are primarily concentrated in the telecommunications sector, which has the highest intensity of technology, while local firms tend to operate in the IT services sector. In this sense, local firms propose themselves as clients of the large foreign operators. The relations between the local firms and the foreign operators can be described as low in knowledge intensity, and firm relations with research centres and universities seem of low relevance. For research and education, DIC seems to depend on the numerous universities of Dubai Knowledge Village, which do not appear to have significant specialisation in ICT (see Fig. 3.2).

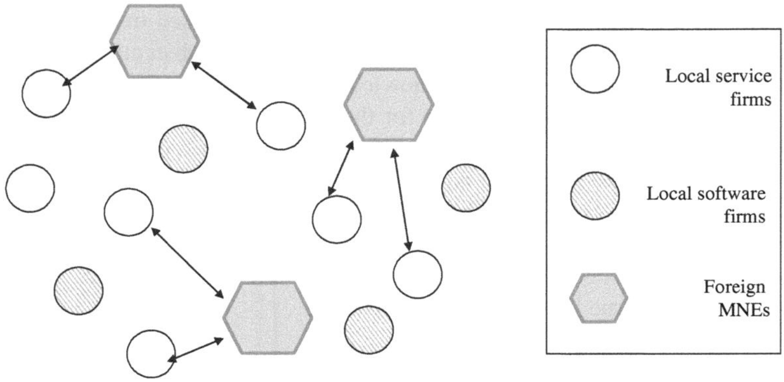

Fig. 3.2 Dubai Internet City

3.4.1.2 Dubai Biotechnology Research Park

The Dubai Biotechnology Research Park (DuBiotech) was officially founded in February 2005 by Sheik Mohammed bin Rashid Al Maktoum, Vice-President of the UAE and Governor and Prime Minister of Dubai, under the framework of the 2005–2010 Strategic Plan. The objective of the park has been to attract the largest global firms in the pharmaceutical and biotechnology sector, to make Dubai the top biotechnology focus in the Middle East. In addition to acting as a focus for firms, the park is also intended as a research centre in the biotechnology and pharmaceutical fields.

The zone is divided in two large areas: one part is intended to host companies while the other is for the research centre. In December 2009, the Nucleotide Laboratory Complex was completed. This is an enormous complex intended to host activities for diagnostics and research, and as a pole of global attraction for researchers in the pharmaceuticals and biological sector.

DuBiotech hosts firms operating in the sectors of pharmaceuticals, medical devices and applied biotechnology for medicine and agriculture, but also a series of service firms active in product distribution for local markets, consulting, events organisation and support for specialised waste disposal. There are currently 55 firms, most of which are foreign. Notable examples are pharmaceutical sector giants such as Pfizer, Merck, Getz Pharma, biotechnology firms such as Genzyme, and medical instruments sector firms such as Philips Respironics and Richard Wolf.

The majority of the local firms are engaged in consulting on marketing, distribution of pharmaceutical products and support for internationalisation activities. These sectors are not high in knowledge content, and local firms have thus served as simple support for the innovative activity undertaken by the multinationals. However, there has recently there been an initiation of gradual knowledge-transfer activity, which has led to emergence of some local operators. An example of this

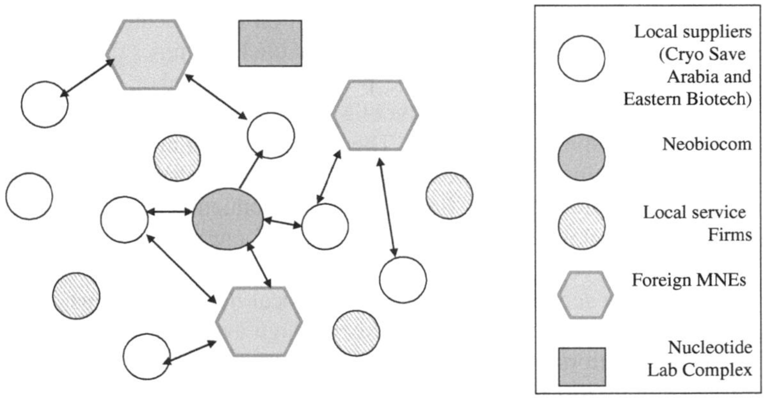

Fig. 3.3 Dubai Biotechnology Research Park

process is provided by the joint venture between the Al Zahrawi Group and Cryo-Save, to form Cryo-Save Arabia, the first stem cell bank in the Arab nations. The firm presents itself a supplier to firms operating in the pharmaceutical and biological sectors, for the stem cells that are fundamental in creation of new pharmaceuticals and for research practice in various fields of biology and medicine. Several years ago, DuBiotech also saw the founding of Neobiocon, the first ever UAE biotechnology firm, as a joint venture between the UAE's Neopharma and India's Biocon. Eastern Biotech is another significant local firm, which carries out testing for identification of various pathologies. An important characteristic is its capacity to deal with the final consumer, an aspect which could serve foreign firms that wish to carry out clinical testing using their proprietary inventions.

However, in spite of the attention given by local government to research activity, including the formation of the Nucleotide Laboratory Complex, there is still no notable presence of important research agencies in the area, and knowledge creation is still restricted to the firms. At the same time, the affirmation of a number of local biotech firms can be considered a signal of the existence of an intense exchange of knowledge among some local operators and foreign firms (see Fig. 3.3).

3.4.1.3 Dubai EnPark

The Energy and Environment Park, or EnPark, was founded as an initiative of Tecom Investment on 7 June 2007, in conjunction with World Environment Day. The business park is a free-trade zone situated on 800,000 ha of land, which presents itself as an advantageous location for firms operation in sectors of renewable energy, bio-construction and environmental consulting. EnPark currently includes 30 firms operating in sectors from creation of infrastructure for development of alternative energy (such as solar panels), consultation in application of sustainable solutions for enterprise, and design of eco-sustainable buildings.

Analysis of the cluster highlights the presence of a number of large international operators, such as Averda and Veolia Environmental Services, which are leaders in waste recycling technologies. There are a limited number of local firms participating in the cluster but most are limited to consulting activity. Notable local firms include Mulk and PTL Solar, active in production of photovoltaic panels, and Pulse Technologies, active in designing and engineering eco-sustainable buildings.

An analysis of the activities carried on by the local firms in the cluster suggests that there are no strong relations between the foreign and local firms. This is partially because the sectors where the local firms operate are complementary to the foreign firms, but not directly related to their technological activities. At the same time, academic activity at EnPark gives particular attention to training for specialised technical personnel, but does not have notable research aspects (see Fig. 3.4).

In conclusion, EnPark currently presents the situation of an agglomeration of firms that do carry out high knowledge-content activity, but which do not have relations among themselves or with research institutions. The cluster assumes a government-centred type of configuration, representing the first stage towards creation of an innovative system.

The analysis of systems of firms in the Dubai area shows how government can use policies of indirect intervention in the economy for purposes of creating local innovation systems. These policies must be designed to increase the absorptive capacity of local firms. As the cases demonstrate, such policies can motivate linkages of foreign firms with local suppliers and stimulate the creation of knowledge spillovers between foreign actors and local firms, and should thus favour the birth of local innovation systems. Although the full desired policy effects are not yet visible, the experiences of the Dubai Biotech Research Park in particular

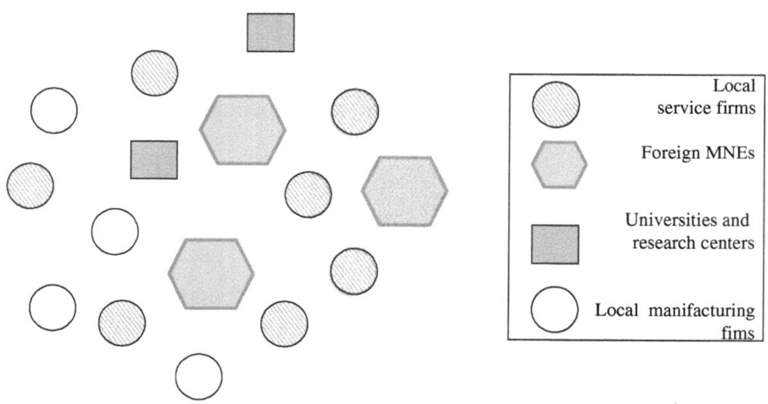

Fig. 3.4 Dubai EnPark

demonstrate that adoption of appropriate government policies can have long-term positive effects, in those cases where they succeed in activating links between local operators and foreign companies, which then favour knowledge transfers.

3.5 Conclusions

The numerous reported experiences of the birth and development local innovation systems in the developed nations have often described a "bottom-up" type of process, where a local large enterprise or university initiates actions for cooperation and knowledge transfer towards other local actors, thus engaging the creation of the LIS.

One of the problems of emerging economies is the absence of local actors, in the form of companies and research centres that would have the size and capacity to activate a bottom-up development process. In these cases, a national government that has interests in reorienting its economy towards specialisation in high-knowledge sectors can intervene, to create the preconditions for development of LISs.

One of the first steps faced in these situations involves the in situ formation of a set of actors that can guide the process of creating an LIS. The process of growing such actors is more complex in situations of lesser development in the local entrepreneurial system. In the absence of any firms or research institutions specialised in a given sector, the birth of an LIS becomes impossible. The state thus has to intervene, guiding the birth process through the creation of the missing actors.

The strategy most frequently adopted is to activate measures intended to convince foreign firms to invest, hoping that knowledge spillovers from these will activate mechanisms for knowledge transfer that can grow local enterprises over time. Governments can implement various policies that not only attract foreign investment but also activate mechanisms that favour the transfer of knowledge from the foreign to local firms.

Government policies can provide for direct or indirect involvement from the state. The Iranian case is an example that appears quite unique in the international panorama for the level of direct government involvement in collaboration with foreign firms. Government is involved through its control over IDRO and the participation of IDRO in joint ventures with foreign firms, and participates in the selection of the foreign partners. The government thus dictates all the conditions for cooperation, including the identification of the knowledge contributions that the foreign firms must provide to local enterprise. In substance, the government and its agencies guide the entire process of knowledge transfer towards local firms.

In the case of Dubai, the government has instead enacted measures intended to attract foreign investment and favour the relations between local actors. The enactment of processes of knowledge transfer then depends on the attitudes of the foreign partners and the capacities of the local actors.

Both the Iranian and Dubai cases show positive results. For example in the Iranian case, the AID-Renault joint venture has afforded numerous advantages to the local partners, who are now able to compete on the both local and international markets. In the second case, the Dubai Biotechnology Research Park is a positive example of how the presence of foreign firms can favour the birth of local high-tech firms.

Both of these national cases show how even in the absence of a local entrepreneurial system, it is possible to engage mechanisms that favour the birth of actors capable of long-term leadership in the creation of an LIS. The Iranian and Dubai cases offer two examples of different policies that governments can implement to engage the formation of local systems, even where the necessary initial elements do not seem to exist.

Obviously these are long processes, and at present only the fundamentals of development have been established in the cases examined. For true functionality there must also be the contemporaneous birth of a research system and creation of cooperation mechanisms between enterprises themselves and between the enterprises and the research institutions.

References

Aadal H, Fard AB, Rad KG (2014) Impact of plants and equipment management in construction industry of Iran. Res J Appl Sci Eng Technol 7(11):2371–2375

Alizadeh P, Hakimian H (eds) (2014) The development of Iran's auto industry in a comparative perspective. In: Iran and the global economy: petro populism, Islam and economic sanctions. Routledge, London, p 157

Austin JW (1990) Managing in developing countries: strategic analysis and operating techniques. Free Press, New York

Blomstrom M, Kokko A (1995) Policies to encourage the inflows of technology through foreign multinationals. World Dev 23(3):159–468

Blomstrom M, Globerman S, Kokko A (1998) The determinants of host country spillovers from foreign direct investment: review and synthesis of the literature. European Institute of Japanese Studies, Working paper 76

Blomstrom M, Sjoholm F (1998) Technology transfer and spillovers: does local participation with multinational matter? National Bureau of Economic Research, Working paper 6816. Available at www.nber.org/papers/w6816

Blomstrom M, Kokko A (2001) Foreign investment and industrial development. Int J Technol Manage 22(5/6):435–455

Brewer TL (1993) Government policies, market imperfections, and foreign direct investment. J Int Bus Stud 24(1):101–120

Buckley PJ, Casson MC (1985) The economic theory of the multinational enterprise. Martin Press, New York

Buckley PJ, Clegg J, Wang C (2007) The impact of foreign ownership, local ownership and industry characteristics on spillover benefits from foreign direct investment in China. Int Bus Rev 16(2):142–158

Cantwell J (1989) Technological innovation and multinational corporations. Basil Blackwell, Oxford

Castellani D, Zanfei A (2003) Technology gaps, absorptive capacity and the impact of inward investments on productivity of European firms. Econ Innov New Technol 12(6):555–576

Caves R (1974) Multinational firms, competition and productivity convergence in host-country markets. Economica 41:176–193

Cheung K, Lin P (2004) Spillover effects of FDI on innovation in China: evidence from the provincial data. China Econ Rev 15(1):25–44
Cohen WM, Levinthal DA (1990) Absorptive capacity: a new perspective on learning and innovation. Adm Sci Q 35:128–142
Crespo N, Fontoura MP (2007) Determinant factors of FDI spillovers: what do we really know? World Dev 35(3):410–425
De Castro JO, Schulze WS (1995) The transfer of technology to less developed countries: a model from the perspective of the technology recipient. J High Technol Manage Res 6(1):113–126
Driffield N, Love JH (2007) Linking FDI motivation and host economy productivity effects: conceptual and empirical analysis. J Int Bus Stud 38(3):460–473
Dunning JH (1993) The competitive advantage of countries and the activities of transnational corporations. Transnatl Corp 2(1):135–168
Dunning JH (1994) Re-evaluating the benefits of foreign direct investment. Transnatl Corp 3(1):23–51
Economist Intelligence Unit (2011) United Arab Emirates country report (April). Economist Group, London
Ethier WJ, Markusen JR (1996) Multinational firms, technology diffusion and trade. J Int Econ 41(1):1–28
Ewers MC, Malecki EJ (2010) Leapfrogging into the knowledge economy: assessing the economic development strategies of the Arab Gulf states. Tijdschrift voor Economische en Sociale Geografie 101(5):494–508
Feinberg SE, Gupta AK (2004) Knowledge spillovers and the assignment of R&D responsibilities for foreign subsidiaries. Strateg Manag J 25(8–9):823–845
Feldman MP (2001) The entrepreneurial event revisited: an examination of new firm formation in the regional context. Ind Corp Change 10:861–891
Ferretti M, Parmentola A (2010) FDI knowledge spillovers and host government policies: the Iranian experience. Eur Bus Rev 22(2):175–194
Filatotchev I, Liu X, Lu J, Wright M (2011) Knowledge spillovers through human mobility across national borders: evidence from Zhongguancun Science Park in China. Res Policy 40(3):453–462
Fosfuri A, Motta M, Ronde T (2001) Foreign direct investment and spillovers through workers mobility. J Int Econ 53:205–222
Ghazinoory S, Divsalar A, Soofi AS (2009) A new definition and framework for the development of a national technology strategy: the case of nanotechnology for Iran. Technol Forecast Soc Chang 76(6):835–848
Globerman S, Shapiro DM (1999) The impact of government policies on foreign direct investment: the Canadian experience. J Int Bus Stud 30(3):513–533
Görg H, Strobl E (2001) Multinational companies and productivity spillovers: a meta-analysis. Econ J 111(475):723–739
Hale G, Long C, (2006) What determines technological spillovers of foreign direct investment? Evidence from China. Yale Economic Growth Center, discussion paper 934, New Haven, CT
Hanson GH (2001) Should countries promote foreign direct investment? G-24 discussion paper series, United Nation conference on trade and development
Javorcik BS (2004) Does foreign direct investment increase the productivity of domestic firms? In search of spillovers through backward linkages. Am Econ Rev 94(3):605–627
Lai M, Peng S, Bao Q (2006) Technology spillover, absorptive capacity and economic growth. China Econ Rev 17:300–320
Lall S (2003) Indicators of the relative importance of IPRs in developing countries. Res Policy 32(9):1657–1680
Lim SH (2005) Foreign investment impact and incentive: a strategic approach to the relationship between the objectives of foreign investment policy and their promotion. Int Bus Rev 14:61–76
Liu X, Wang C (2003) Does foreign direct investments facilitate technological progress? Evidence from Chinese industries. Res Policy 32:945–953
Liu X, Buck T (2007) Innovation performance and channels for international technology spillovers: evidence from Chinese high-tech industries. Res Policy 36(3):355–366
Markusen GR, Venables AJ (1999) Foreign direct investment as a catalyst for industrial development. Eur Econ Rev 43(2):335–356

Markusen JR (2002) Multinational firms and the theory of international trade. MIT Press, Cambridge
Meyer KE, Vo Nguyen H (2005) Foreign investment strategies and sub-national institutions in emerging markets: evidence from Vietnam. J Manage Stud 42:63–93
Meyer EK (2004) Perspectives on multinational enterprises in emerging economies. J Int Bus Stud 35(4):259–276
Mirzaei R (2012) A review of successful housing system in Iran and the introduction of cob as a new alternative in Iran's industry. J Am Sci 8(10):180–184
Narula R, Dunning JH (1998) Explaining international R&D alliances and the role of governments. Int Bus Rev 7:377–397
Nunnenkamp P (2004) To what extent can foreign direct investment help achieve international development goals? World Econ 27(5):657–677
Pietrobelli C (1996) Emerging forms of technological cooperation: the case for technology partnerships—inner logic, examples and enabling environment. Science and technology issues. UNCTAD, Geneva
Porter ME, Monitor Group, on the FRONTIER, Council on Competitiveness (2001) Clusters of innovation: regional foundations of US competitiveness. Council on Competitiveness, Washington DC
Rodriguez-Clare A (1996) Multinationals, linkages and economic development. Am Econ Rev 86(4):852–873
Saggi K (2006) Foreign direct investment, linkages and technology spillovers. In: Hoekman B, Javorcik BS (eds) Global integration and technology transfer. Palgrave Macmillan/World Bank, Washington DC, pp 51–65
Singh H, Jun KW (1995) Some new evidence on determinants of foreign direct investment in developing countries. Policy research working paper, World Bank International Economics Department
Sjoerd B, Smeets R, Zwinkels R (2008) The impact of horizontal and vertical FDI on host's country economic growth. Int Bus Rev 17(4):452–472
Smarzynska BK (2002) Does foreign direct investment increase the productivity of domestic firms? In search of spillovers through backward linkages. World Bank policy research, Working paper 2923, 2002
Smeets R (2008) Collecting the pieces of the FDI knowledge spillover puzzle. World Bank Res Obs 23(2):107–138
Te Velde DW (2001) Policies towards foreign direct investment in developing countries: emerging best-practices and outstanding issues. Available at www.odi.org.uk
Te Velde DW (2002) Government policies towards foreign direct investment. In: Wignaraja G (ed) Competitiveness strategy in developing countries: a manual for policy analysis. Routledge, London
UNCTAD (2005a) World investment report 2005: TNCs and the internationalization of R&D. United Nations conference on trade and development, Geneva
UNCTAD (2005b) World investment report 2003–2005. United Nations Conference on Trade and Development, New York
Wang JY, Blomstrom M (1992) Foreign direct investment and technology transfer. Eur Econ Rev 36:137–155
Zhang Y, Li H, Li Y, Zhou LA (2010) FDI spillovers in an emerging market: the role of foreign firms' country origin diversity and domestic firms' absorptive capacity. Strateg Manag J 31(9):969–989

Chapter 4
The Firm-Driven LISs

Abstract A substantial body of literature explains how the presence of a leading firm in an LIS has positive effects on cluster development. A number of studies underline the effects that leading organizations can have on spillovers, particularly in facilitating knowledge circulation and avoiding lock-in phenomenon. Such potentials exist especially for large firms. This chapter analyses the mechanisms that leading firms can implement to favour the formation and growth of local systems. It then provides a case study of the role of Acer Inc. in developing an information-technology LIS centred around Hsinchu, Taiwan, illustrating the positive effects from a leading firm.

Keywords LIS · Leading firms · Spillovers · Taiwan · Hsinchu cluster · Acer

4.1 The Role of the Large Firm in Development of the LIS

According to mainstream literature, learning and innovation in the business sphere are the result of interactive processes in which different actors come together to collaborate in solving specific problems. Many scholars have emphasised how important it is to be part of a local innovation system, in order to have favourable conditions to trigger the innovation process at both the firm and cluster levels.

Researchers have demonstrated that the participation of firms in an LIS permits important localised knowledge spillovers. The firms benefit from the availability of a pool of skilled labour, and given their geographical and social proximity, new ideas circulate easily from one business to another, promoting processes of incremental and collective innovation (see among many others: Becattini 1989; Asheim 1994; Saxenian 1991; Audretsch and Feldman 1996; Maskell and Malmberg 1999; Baptista 2000; Parilli 2009).

Other researchers have stressed the importance of extra-cluster networking. This is because overreliance on local knowledge alone can result in the "entropic death" of the LIS, as it remains locked in an increasingly obsolete technological

M. Ferretti and A. Parmentola, *The Creation of Local Innovation Systems in Emerging Countries*, SpringerBriefs in Regional Science,
DOI 10.1007/978-3-319-10440-9_4

trajectory (Camagni 1991; Grabher 1993; Cantwell and Iammarino 2003; Bathlet 2004; Dei Ottati 2009).

The search to understand the nature and effects of spillovers generated within a local system has long dominated the attention of scholars. In particular, several studies have underlined the positive effects that the presence of leading organizations, especially large firms, can have on spillovers, in facilitating knowledge circulation and avoiding lock-in phenomenon (Boari and Lipparini 1999; Lazerson and Lorenzoni 1999; Agrawal and Cockburn 2003; Feldman 2003). The lead firm favours the diffusion of knowledge among LIS members and acts as an interface to absorb new knowledge from external sources (Munari et al. 2011).

A number of contributions explain how the presence of a leading firm in an LIS positively affects cluster development. Some scholars have conducted detailed analysis of the evolution of industrial districts, illustrating the pivotal role played by key actors and institutions (Boari and Lipparini 1999; Dhanaraj and Parkhe 2006). Lorenzoni and Baden-Fuller (1995) focus on the role of what they call "focal firms" within industrial districts. These are defined as "strategic centres" that can assure the survival and development of the entire district, thanks to their superior coordinating skills and ability to steer other firms toward innovation cycles and new growth opportunities. These actors are key to generating new knowledge and technologies, in spinning off innovative companies, attracting researchers, investments and research facilities, in enhancing other firms' R&D activities, stimulating demand for new knowledge, and in capturing and creating externalities.

Studies of several Italian industrial districts show how large firms have played a fundamental role in collective learning processes, acting as drivers for innovation development and cluster growth (Boari and Lipparini 1999; Lazerson and Lorenzoni 1999). In particular, Lissoni (2001) illustrated how the Brescia mechanical district (northern Italy) was quite dependent on a small number of major firms with global markets, such as Lonati Spa, to foster incremental innovation and the welfare of the whole district. Another example of leading firms' centrality in district development is that of Benetton, which developed many relationships with smaller producers and distributors to manage subcontracting and outsourcing of production, thus stimulating efficiency and innovation development among the cooperating firms (Camuffo et al. 2001).

Empirical studies demonstrating the role of leading firms in local economic growth and innovation development are not restricted to Italy. Analysis of the Chilean wine cluster (Giuliani and Bell 2005) shows that more advanced firms tend to diffuse knowledge and provide advice to other firms located in the cluster, thus promoting positive externalities for the whole system. In the case of watch production in Switzerland, the major firm ETA SA Swiss Watch was largely responsible for the re-founding of the industry, thanks to the development of digital technologies that in turn led to the birth of Swatch (Glasmeier 1991). The Scandinavian LIS of wireless hardware firms benefitted greatly from the phenomenal growth of two major companies, Eriksson and Nokia (Breshanan et al. 2001; Van Winden et al. 2004). There are similar examples of the role of key firms in development of LISs in the United States, such as the well-known cases of Fairchild Semiconductor and Intel in Silicon Valley (Utterback 1994).

The literature has specifically identified a series of features that characterise such focal firms. Lazerson and Lorenzoni (1999: 362) define "focal" as "those firms that occupy strategically central positions because of their greater number and intensity of relationships with both customers and suppliers. This position is usually reinforced by both their technological and organizational skills and their greater access to capital." Similarly, Giuliani and Bell (2005) emphasise the superior stock of knowledge accumulated by such firms as a result of their intense R&D and their greater assets of skilled human resources.

Analysing the innovation process in regional innovation systems, Carbonara (2004) states that the presence of a lead firm in a cluster helps systemic innovation by favouring knowledge acquisition and diffusion processes among different actors. Lead firms make intense efforts to codify knowledge, especially technical knowledge, so that it is easier to control and transfer. This in turn helps continuous innovation, permits reduced lead times, and improves product quality and communication with external actors (Albino et al. 1999). Consistent with this argument, some authors affirm that leading firms can act as "technological gatekeepers" in an LIS, by leveraging their intellectual and social capital to enhance the absorption of new information and facilitate its dissemination (Munari et al. 2011; Camuffo and Grandinetti 2011). As gatekeeper, the leading firm carries out (Morrison 2008, p. 8):

- knowledge searching functions, for capturing external sources of information;
- transcoding functions, for translating the meaning of such information;
- transferring functions, for disseminating accumulated knowledge at the local level.

Technological gatekeepers are characterised by the ability to diffuse re-elaborated knowledge to others, through a process that can be deliberate or not. Thus, focal firms can play a distinctive role in the diffusion of new knowledge from outside sources to other firms co-located in the system.

Some scholars use the term "anchor tenant", similar to the concept of the technological gatekeeper, to emphasise the important function of leading firms in favouring LIS growth.

There are a number of studies that explicitly examine the concept of the anchor tenant and offer input towards a rigorous definition. In their studies of the biotechnology industry, Feldman and others (Feldman 2003; Feldman and Lowe 2008) define anchor firms as large, technologically sophisticated entities creating externalities (e.g. a pool of skilled labour, demand for specialised inputs) which could benefit smaller local firms. Agrawal and Cockburn (2003: 1,229) define an anchor tenant as a large, locally present firm that is heavily engaged in extensive R&D. Link et al. (2003) identify high-technology anchor tenants as large R&D-intensive firms, as recognised by their patenting activity, that have a strong focus on a particular technological field. They argue that high-technology anchor tenants enhance local systems by stimulating technological externalities through their own actions. Niosi and Zhegu (2005) applied the concept of anchor tenant in analysing the role of large aerospace firms in regional innovation systems of the Canadian aircraft industry. These companies have most often started such clusters by favouring processes of agglomeration among other smaller firms in the area. In a more recent work, the same authors (Niosi and Zhegu 2010) collect empirical evidence for a series of aircraft clusters in the United States, and show that some have failed because the leading firms transferred their activities to new geographic locations.

4.1.1 Mechanisms Used by Large Firms to Grow the LIS

The heterogeneity of the actors in an LIS represents a critical issue in achieving economic development from any research results. This is particularly true where the different cultural orientations of the organisations create barriers to relationships. Overcoming these barriers requires that certain common values take root among the universities, firms and local institutions of an LIS, in order to orient the behaviours of the different actors towards shared trajectories of development. It is particularly necessary that the universities and research centres interiorise the logic of economic exploitation of scientific discoveries and technological innovation, overcoming traditional attitudes that might identify the role of research organisations as exclusively in extending the frontiers of knowledge. On the other hand, it is indispensable that the firms also recognise the working procedures and times involved in scientific research, as a reference for their implementation of projects involving structures or individuals from the academic sphere. Finally, public institutions must acquire more specific and deeper knowledge of the situations and needs involved in both the research and the business innovation spheres, in order to orient public interventions in a coherent manner. However it is not easy to enact the process of sharing a set of central cultural values, necessary for the systematic action of the innovative hub. The specificity of the different types of actors, the increasing rigidity and sedimentation of organisational cultures over time, and the inertia of individual narratives and past successes, will all present obstacles to the hybridisation of the various individual identities with the shared values necessary for the entire LIS. In attempting such cultural diffusion, the large firm can play an important role. Through their dense relational networks, such firms are in an optimal position to extend their own internal innovation management models through the entire LIS.

The large firm can draw on different mechanisms to create a shared culture within the LIS, particularly in the relations between academic and managerial actors. A first type of mechanism consists of inserting individual managers in the teaching and research activities organised by the universities and research institutions. Initiatives of this kind, repeated over time, permit the managers to establish personal contacts with research officers and faculty members, through which other sharing opportunities can be organised.

A second type of mechanism available for the large firm to act as culture diffuser in the LIS is through "board interlock". Understood here in the broad sense, "interlocks" are the links that form between organisations when the same person holds a formal position in the different entities, most typically when the individual sits on the board of directors of different companies. The capacity of interlock to act as a vector for managerial and cultural models arises from the person's knowledge of the different entities in which he or she works (Salancik and Pfeffer 1978). For large firms in the LIS context, there are several possible forms of interlocks. For example, a manager could be assigned to an important position within the organisation of another actor in the LIS, or the president of a large company could be co-opted to the board of directors of an important university in the territory.

A third type of mechanism for the large firm to culturally orient the other organisations in the LIS is through the role of its managers in the meta-organisational structures and activities, created for the purposes of the hub itself.

The large firm of an LIS can also play an important role as facilitator in the management of specific interdependencies between the different actors in the system. Through the non-binding links that the large firm managers are able to establish with the other LIS actors, they can contribute to easing the relationships between the communities of researchers, managers and financiers, in various ways.

The first method involves cooperative relationships, based on handshake agreements, through which the managers can act as guarantors for the validity of a new entrepreneurial initiative. The presence of a link with the large firm, guaranteed by the involvement of one of its top managers, can in fact heighten perceptions of the validity of a business plan developed by the promoters of a particular project. In an LIS environment, for example, top managers of a large firm could encounter possibilities to support the initial stages of an academic spin-off, through a handshake agreement. Or there could be instances of informal cooperation by large firm managers in applied research projects, aimed at patenting innovations. In all these cases, even without formal links, the simple involvement of the manager eases the dialogue and eventually the collaboration between researchers interested in developing a new technology, the entrepreneurial talent interested in economic exploitation of the idea, and the potential investors in the financing of the project.

A second mode for the large firms' links to contribute to interdependencies between ideas, talent and capital is through the mobility of the individuals. The movement of a person from the large firm to another organisation within the LIS brings out two important effects. First, it generates a spillover of technological and market knowledge that can increase the potential for future economic development of research results (Simmie 2002). Second, the passage of a manager from a large firm to another organisation in the LIS reinforces the image of the smaller one. The personal curriculum and reputation connected to the individual's work in the large firm become assets for the new employing organisation. Hence the latter, in an indirect manner, increases its credibility with respect to the relevant communities of researchers, entrepreneurial talent and financiers.

The presence of a large firm in an LIS will most certainly permit at least partial resolution of the problems of under-sizing, which are typical of many activities in the initial phases of the system development. The large firm can act as supplier of critical resources for different typologies of actors and instruments. Concerning the technological platforms, the large firm can contribute to development of major infrastructure and research laboratories, which the other LIS actors would not otherwise be able to create. Or as an alternative it could provide the universities and research centres with access to its own laboratories, where these are already equipped with advanced and costly technologies. Concerning the labour market, the large firm can serve as an actor in the training of qualified personnel, through its own structures. Some of the professional profiles necessary for the different initiatives in promoting business development of research results can in fact be developed within the large firm, whether through learning by doing or by more classical forms of education. Such training could be in areas such as the managerial competencies necessary for new firms, or the technological competencies for development in new fields. Thus the large firm can in part resolve the local lack of offer in the management and technical competencies necessary to feed the development projects of the different actors in the system.

Finally, concerning the problem of the necessary capital for initiatives in the business development of research results, the large firm can intervene as a financier. This type of intervention generally takes place in one of two modes. The first is direct intervention by the firm in a project for technological transfer or creation of a new company. In this case, the large firm invests part of its available resources in an initiative from which it believes it can obtain long-term advantages. The second mode is through formation of a corporate venture capital fund. In this case the firm allocates its resources for the creation of the investment fund, with then has operational autonomy. Once the mandate for investment in specific fields has been established, the autonomous structure permits the fund managers freedom to select projects under conventions different from those of the large firm that provided the capital.

Interventions are particularly necessary in the early creation stages of an LIS, in order to facilitate the initiation of economic development processes for research results, and their progressive consolidation and strengthening. The nascent innovative hub must be sheltered from events that could prejudice its survival during the stages of the first scientific and technological innovations, especially when university research activities are not fully developed, the actors in the system are still few, and when strong relations have not consolidated and even weak ones are still rare (Amin and Cohendet 2000). External factors could reduce the growth possibilities for the entire system: competitive pressure from other territories involved in research in analogous domains; structural changes in the potential application sectors for the research being conducted by the LIS universities and research centres; repeated changes in the economic policies of government institutions, particularly if in directions unfavourable to the territorial actors. The large firm can offer valid responses to the needs for protection of systems in their early development stages, presenting itself as the primary intervener both for the universities and the firms present in the territorial system, formed with intentions to exploit the newly developed technologies. The large firm can implement agreements to support demand, for both the new technologies developed by the research centres and the products and services offered by the firms. Such actions render the local dynamics relatively independent of events external to the territory (Wolfe and Gertler 2004). The large firm can act as the buyer for applied research projects, outsourcing them to the universities. In this way, the firm can support the research efforts of the LIS actors, permitting the various figures to accumulate the knowledge necessary to proceed along the path of developing the entire system. Concerning products and services, the large firm can guarantee a market outlet for the fabric of small, innovative firms, absorbing their offer and utilising it in the realm of its own business activities (Boari and Lipparini 1999; Agrawal and Cockburn 2003). In this way, the small firms can be relatively protected from competitive pressures, potentially generated by analogous innovative firms present in other territories, and begin to grow in size.

4.2 The Role of Acer Inc. in the Hsinchu, Taiwan LIS

Hsinchu is a small administrative district of Taiwan (Republic of China), less than 100 km from Taipei. Today this area is considered as the "Asian Silicon Valley" (Chen 2008; Chang 2009) and the symbol of the national economic miracle, having elevated Taiwan to the level of a global leader in the sector of personal computer semiconductors. Whereas the nation was formerly known for its economy specialised in mature production forms with low added value, it now ranks as the fourth in the world for number of patents registered annually.[1]

The Hsinchu local initiative system currently counts 384 high-tech firms, two avant-garde universities and a major public research institute. These latter institutions are the National Chiao Tung University, the National Tsing Hua University and the Industrial Technology Research Institute (ITRI). Alongside the myriad of small companies that compose the system there are several large companies, such as Acer Inc., which at a certain point took a guiding role in the management of relations between the various actors in the cluster. It is thanks to the presence of these firms that the Hsinchu area has evolved over time, asserting itself as one of the most advanced technological hubs on the world scene.

It is difficult to precisely identify the spark that ignited the LIS formation process, however it seems clear that the preconditions that led to the birth of the Hsinchu system date back to the 1970s, when the first Taiwanese ICT firms were formed. In fact in roughly those years, many Taiwanese engineers with career experience in Silicon Valley decided to return to their homeland, and found their own companies that could serve in low-cost product assembly for the large American firms (Saxenian and Hsu 2001; Chyi et al. 2012).

At the same time, the government, well aware of the opportunities that development in this sector could bring to the country, inaugurated the Industrial

---

[1] The United States Patent and Trademark Office databases shows that as of 2008, Taiwan (Republic of China) placed fourth in the world for number of patents, after the United States, Japan and Germany, and that Taiwan was first among Asian nations.

Technology Research Institute (ITRI). This institution co-opted the top Taiwanese and foreign researchers for development of a centre of research excellence (Hung and Tang 2008). With the aim of favouring innovation, the government also incentivized the formation of numerous consortiums between the ITRI and local firms, and encouraged patenting of innovative solutions in the ICT field (Mathews 2001).

Several years after these opening events, a number of foreign multinationals, such as RCA, Texas Instruments and Philips, recognised the ferment in the area and opened branches in Hsinchu. In this way they were able to exploit the newly qualified manpower at low cost (Hobday 1995).

However the true turning point took place in the 1980s, in connection with the development of two important events: the birth of the Hsinchu Science Park and the consolidation of relationships with Silicon Valley (Yang et al. 2009).

The decision to create a scientific and technological park in the Hsinchu area was not at all casual. The choice of this location was dictated by a set of conditions. Hsinchu County was the area where ITRI had its main laboratory, and also the seat of the country's biggest university. The large part of the firms that had developed in the 1970s were based in Hsinchu or adjacent areas in northern Taiwan. And finally, the area was characterised by a good presence of infrastructure, due to its proximity to capital city of Taipei (Hu et al. 2005).

With the creation of the science and technology park accomplished, the government next established a series of tax-related incentives intended to attract firms to the area. This in turn stimulated a number of firms operating in other national regions to also move their activities to Hsinchu, again in the 1980s (Chen 2008). The whole of these events created the aspect of spatial proximity between different actors, which favoured the birth of spillover phenomena.

In these same years, the informal links that had developed between the repatriating Taiwanese engineers, the local community and the American branch firms began to consolidate. From this, true longer-term supply agreements began to emerge. In fact the Silicon Valley firms began to externalise low added-value, assembly-line activity to the Taiwanese companies. The multinationals planned the product and transferred the detailed fabrication plans to their assemblers. These proceeded with the production of the simpler components and the final assembly, then returned the finished product to the mother firm, which provided the distribution to the end customers (Wong 1999). The process permitted the local firms to acquire the necessary competencies for operations in the personal computer sector. They were able to rapidly fill the gaps in the competencies and activities of the local research centres, still quite far from the leading edge on the theme, while the contacts with international distribution systems compensated for the lack of significant local demand for ICT products.[2]

---

[2] As amply discussed in Chap. 2, two of the conditions necessary for the birth of an innovation cluster are the presence of advanced research centres, for acquisition of competencies, and of end demand that stimulates the creation of innovative products. In many emerging nations these elements are not present, but they can be effectively substituted by relations with foreign multinationals and with technology clusters that already exist in more advanced nations.

However in subsequent years, the personal computer sector underwent product evolution. Many large American firms abandoned the sector of the desktop p.c., refocusing on what were then more profitable sectors in notebooks, netbooks and palm devices.[3] Others transferred their production to countries offering manpower costs that were still lower, such as mainland China. At the same time, the tendencies towards miniaturisation of components and greater emphasis on the aesthetic design of products brought about necessities for continuous innovation in the companies that were still operating in the sector. Such innovation often required research in a series of related fields, for example in nanotechnology.

All of this brought about an abrupt re-dimensioning in the intensity and typology of links between the Taiwanese firms and the American multinationals. In these years the Hsinchu cluster would certainly have disappeared were it not for the local firms' success in achieving technological independence from the American companies, through investing in R&D activities and cooperating with the local research centre and universities.

In fact, beginning in the 1990s, with the ever greater commoditization of computer products and the simultaneous birth of new market niches, the Taiwanese companies began to acquire autonomous capacities in planning and manufacture of new products. In these years, some companies continued to work for the American firms, carrying out not only low value-added activities, but also pushing into the level of planning the entire production. The American firm Compaq, for example, continued to externalise its production to Taiwan, however the local firms were no longer limited to carrying out activities of lower knowledge content. Instead they began to produce the entire computer, from the initial design and planning up to the construction of the complete product, which was then commercialised under the Compaq brand. This was a case of local firms shifting from an "original equipment manufacturing" to "original design manufacturing" stage (Hobday 1995).

Other firms would instead choose the road of specialising their production in the realisation of specific components, such as hard drives and DVD recorders.

With the birth of the so-called "clone computers", some producers then specialised in the realisation of low-cost products, meaning computers that reproduced the entirety of characteristics of the more costly IBM types, but which were commercialised at markedly lower prices. Among other considerations, this strategy achieved products that were particularly adapted for commercialisation in emerging markets (Ku et al. 2005).

Other firms, most notably Acer, through investing in innovation and marketing, became true ICT multinationals, capable of competing on equal footing with the American giants. The birth of this dimension of firm also led to a change in the relationships within the system. The new multinationals began to dialogue with the existing research centres and the university, offering these latter institutions a constant stimulus to continued research in the semiconductor and personal computer

[3] An example is IBM, which ceded almost its entire personal computer division to the mainland Chinese company, Lenovo.

sectors. They also substituted the foreign firms in giving work to the other firms in the cluster, creating true sub-supplier networks, and originated new firms through spin-off processes. In other words, they became the central axis around which the entire innovative system rotated.

Thus, beginning in the 1990s, the Hsinchu LIS ceased to be a satellite cluster of Silicon Valley and established itself as a true LIS, characterised by the presence of numerous large and small actors, with their actions coordinated by a single large, central actor, in the form of Acer. From this moment, the story of the Hsinchu system essentially becomes the story of Acer.

4.2.1 The Actors and Their Relations

To continue the study of the Hsinchu LIS we must first briefly examine the range of actors involved. For this we apply a typical classification of the actors in local innovation systems (Etzkowitz and Leydesdorff 2000) to analyse the roles of the three fundamental types of organisations in the cluster: the university and research centres, government institutions and the firms. It is well recognised that while development of an innovation system requires the co-existence of these three types of actors, it depends even more on the relationships that they are able to develop (Edquist 2005). The analysis thus highlights the contribution that each component has brought to the system, from a relational point of view (Fig. 4.1).

4.2.1.1 The Hsinchu Universities and Research Centre

As noted above, the choice of the Hsinchu area as location for the country's largest scientific and technological park was partially dictated by the presence of two of Taiwan's largest state universities, the National Chiao Tung University (NCTU) and the National Tsing Hua University (NTHU), as well as the Industrial Technology Research Institute.

National Chiao Tung University

The NCTU is the oldest university in the country, originally founded at Shanghai in 1896 and the re-founded in the Hsinchu area in 1958. The institution is at the vanguard of research in the engineering and electronics sectors. It places among world leaders for number of publications in international scientific journals. The large part of the founders and the engineers employed in Hsinchu LIS firms received their advanced education from this university. The links between alumni are strong, with the institution's annual reunions heavily attended, often serving as the occasion for knowledge exchanges between managers and the development of new inter-company agreements.

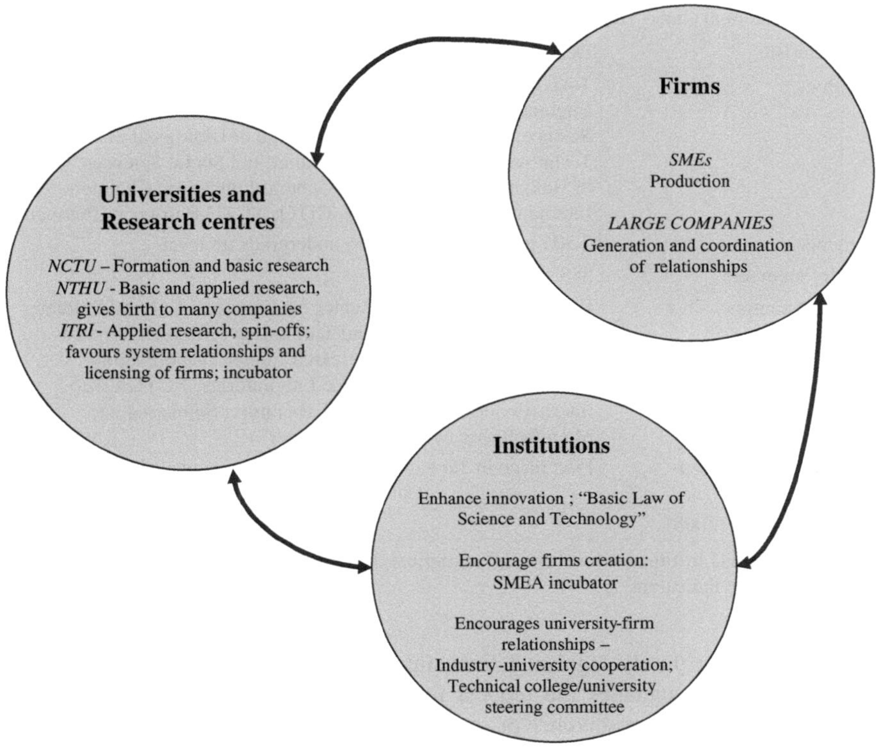

Fig. 4.1 Actors and roles in the Hsinchu LIS (from Parmentola 2009)

In addition to its educational function, the university is at the vanguard in scientific research in service to private enterprise. Only a few years after its re-establishment in Taiwan, in agreement with the Ministry of Education and with special funding from the United Nations, the university founded the Telecommunication and Electronics Training Center (TETC). TETC introduced computer technology to Taiwan, initiated television broadcasting, and manufactured the island's first transistors and solid-state lasers. After this the university established a series of further research centres. In 1964, again at the request of the central government, but now also former students, the university's departments of engineering and electrophysics founded the Electronics Institute. In 1967 this institute was formally reorganised as the College of Engineering, composed of a series of departments. This particular college is still the pride of the entire university.

The NCTU, like other Taiwanese universities, has an exceptionally organised practice concerning management of patents registered by its researchers and laboratories. In 1996 the university founded the Office of Research and Development, responsible for the commercialisation of the university's research results, the concession of licenses for patents achieved by the university, and the management of grants and other funds awarded by the companies of former students.

Table 4.1 National Chiao Tung University—characteristics

Characteristic	Description
Faculties	Total of 9: College of Engineering; College of Electrical Engineering; College of Management; College of Computer Science; College of Science; College of Biological Science and Technology; College of Humanities and Social Sciences; College of Hakka Studies; Center of General Education; Chalmers International Taiwan Office, CITO; NCTU Europe at Chalmers
Students	7,505 graduate level; 4,786 undergraduate level
Faculty members	588
Research centres	Brain Research Center; Center for Emergent Cultural Studies; Center for Nanoscience and Technology; Microelectronics & Information Systems Research Center; Nano Facility Center; National Nano-device Laboratories; NCTU/Cisco Internetworking Technology Laboratory; Semiconductor Manufacturing Management Center
ESI[a] papers published (2008)	First place in Taiwan (Computer Science category)
Patents obtained (2008)	8

Data source NCTU institutional website, all numerical data as of 2008
[a]Essential science indicators

Other than the purely legal and financial aspects, the Office of Research and Development also manages the request for patent registrations advanced by the university faculty. The researcher or research group that achieved the invention is required to first submit the patent proposal to a committee of faculty members that evaluates the research quality. If the request passes this step, the patent proposal is then submitted to a group of external reviewers. Only then is the actual patent registration submitted, under the name of the university.

The relatively small number of patents registered by the university in a typical year (Table 4.1) indicates how the NCTU concentrates its contributions to the LIS on the specific functions of educating highly qualified engineers and managers, as well as in investment for basic research. The emphasis on basic rather than applied research is also revealed by the high number of scientific publications achieved by the faculty and staff.

4.2.1.2 The National Tsing Hua University

The Tsing Hua University was originally formed in 1911 in Beijing. In 1955, academics and staff who had moved from the mainland founded the National Tsing Hua Institute of Nuclear Technology, which one year later became the National Tsing Hua University. The university has constantly maintained the status of a centre of excellence in the fields of physics, chemistry and electronics. Currently it counts seven faculties, 17 departments and 19 independent research institutions. The university alumni include Cheng-Ning Yang and Tsung-Dao Lee, both

Table 4.2 National Tsing Hua University—characteristics

Characteristic	Description
Faculties	7 in total: College of Science; College of Engineering; College of Electrical Engineering and Computer Science; College of Life Sciences; College of Humanities and Social Sciences; College of Nuclear Science; College of Technology Management; Commission of General Education
Students	5,194 graduate level; 5,071 undergraduate level
Faculty members	599
Research centres	19 in total: Research & Development Office; Computer & Communication Research Center; National Center for Theoretical Sciences; IC Design Tech. Center; Center for Science Technology and Society; Center for Nanotechnology, Materials Science, and Microsystems; Center for Photonics Research; NTHU/ITRI Joint Research Center; Center for Contemporary China; Center for Energy and Environmental Research; Brain Research Center; Laboratory Animal Room; Instrument Center; Nuclear Science and Technology Development Center; Electronic Business Center; Quality Research Center; Advanced Packaging Research Center; Bioinformatics Center
ESI[a] papers published (2008)	Approximately 1,000
Patents obtained (2008)	56
Innovation incubators	One: Open Lab Innovation Incubation Center

Data NCTU institutional website, all numerical data as of 2008
[a]Essential science indicators

Nobel prize winners in physics, and Yuan-Tseh Lee, winner of the Nobel prize for chemistry, as well as Shiing-Shen Chern, winner of the Wolf Prize in Mathematics.

A review of the university's principle characteristics (Table 4.2) quickly reveals that the NTHU is more focused than its sister university on basic and applied research, rather than education. This fact is illustrated by the impressive number of patents registered, at over 50 per year, and the number of articles published in international journals recognised by the Science Citation Index, at roughly 1,000 per year.

In 1998 the university created an Office of Technology Service and Licensing, to achieve more efficient management of procedures for patent applications by university personnel. The office also assumed general responsibility for management of intellectual property. The authorisation procedure for patent submission under the university name follows two evaluation stages, similar to the NCTU provisions.

The NTHU has greater entrepreneurial propensities compared to the NCTU. In 1998 it created an innovation incubation centre, with the stated aims of assisting and stimulating the creation of new firms, favouring cooperation between the university and enterprises, and reducing risks and costs by accompanying entrepreneurs in the creation of new companies. As of today the incubator has supported the birth of 34 companies, 85 % of which have become part of the Hsinchu LIS. Another 24 projects are currently in incubation.

The analysis of the NTHU policies, as well as the evidence of the high number of patents, reveals that the contribution of this university to the direct evolution of the system is much greater compared than that offered by the NCTU. The university not only provides education for the personnel employed in the companies, but its orientation towards applied research also permits the creation of innovations, which can then be conceded under license to the local firms. Finally, the incubation activity favours the continued renewal of the cluster through the creation of new innovative companies.

4.2.1.3 Industrial Technology Research Institute

ITRI was founded in 1973 by the Taiwan Ministry of Economic Affairs, through the merger of three public research organisations: the Mining Research & Service Organization, the Metal Industrial Research Institute, and Union Industrial Research Laboratories. The creation of ITRI permitted the Taiwanese economy to make remarkable progress, initiating the development process that led from an economy based on high-manpower sectors to a new basis in high technology. In fact in 1976, ITRI initiated the semiconductor sector in Taiwan, signing an agreement with the American firm, RCA, for the acquisition of the CMOS technology that is the basis of semiconductor production (Chen and Choi 2004).[4]

Today, ITRI still contributes to the development of the Hsinchu system through a variety of means, including via the contributions of the 6,200 researchers employed in its research laboratories. Unlike the situation for the two universities in the LIS, ITRI's activity is strongly oriented towards applied research, and specifically in the key sectors of the Taiwan economy: semiconductors, liquid crystal screens, nanotechnology, broad-band mobile communications, biomedicine and bio-informatics. This specialisation is demonstrated by the roughly 3,000 patents per year that ITRI registers with the United States Patent Office, concentrated primarily in sectors G and H (Physics and Electricity), which correspond to Taiwan's strategic economic sectors.

ITRI ensures the acquisition of technological knowledge not only through its internal research, but also by stipulating cooperation agreements with foreign firms and research institutions. For example in 1999, ITRI reached an agreement with Alcatel, the French multinational, for an ADSL Interoperability Laboratory, with the mission of testing the efficiency of ADSL communications technologies and planning new solutions. In 2004, ITRI established DelSolar Co., a joint venture with Delta Electronics, which became the top producer of photovoltaic

[4] The complementary metal-oxide semiconductor (CMOS) is one of the base technologies for integrated circuits, used for microprocessors, microcontrollers, static memories and other types of digital logic circuits. The technology was patented by Frank Wanlass in 1963.

cells in Asia. In 2003, ITRI instituted a cooperation programme with the Stanford Research Institute, the Carnegie Mellon University and Berkeley University, and reinforced existing cooperation agreements with MIT, Northwestern University in Illinois, and the Tohoku University of Japan.

The Industrial Technology Research Institute also plays a key role in activities for diffusing innovation, above all through awarding licenses under its numerous patents. To facilitate this process it created an information platform called Technology Mart, which functions as a sort of on-line auction for the sale of patent licenses. The service offers a virtual meeting point for a range of research centres and the firms that are the potential buyers, who interact by posting, perusing and responding to patent descriptions online. The bidding for the individual licenses is conducted on a weekly basis, with the highest offer receiving the license. This mechanism favours both the firms, which are constantly informed of new innovations and can purchase them soon after development, and the research centres, which can maximise income from licensing their patents.

ITRI also favours the diffusion of innovation by encouraging its researchers to establish their own companies. The first ITRI spinoff, founded in 1979, was United Microelectronics Corporation (UMC), Taiwan's first semiconductor company and the first firm in the nation to use the 4-in. wafer technology.[5] Another firm that traces its origins to ITRI is TSMC, founded in 1987, the first Taiwanese company to produce 6-in. wafers, and now the third-ranking semiconductor firm in the world. In 1994, ITRI budded Vanguard International Semiconductor Corp., the first company in Taiwan to produce processors using 8-in. wafers, followed in 2003 by the Phalanx Biotech Group, which operates in the biochip field. Many other companies born as ITRI spin-offs are now at the heart of the Hsinchu cluster.

ITRI also favours the creation of firms by financing innovations and ideas with maximum profit potential through its Open Lab project, an incubator founded in 1998, which assists in the start-up stage of highly innovative firms.[6] As of the current date, Open Lab has supported the birth of 86 companies. The large part of these operate at Hsinchu, thus composing nearly half of the total cluster. They include Phison Electronics Corp. (the world's first producer of the single-chip USB flash drive), Prolific Technology (microchip producer) and Taiflex Scientific Co. (semiconductor materials manufacturing).

But ITRI's role is not limited exclusively to creation and diffusion of innovation. The institute has also always served as an education facility, and in 2003 founded ITRI College, which offers courses in science and technology,

[5] The wafer is a thin slice of semiconductor material, such as a silicon crystal, used in the fabrication of integrated circuits and solar cells. The wafers are available in a variety of formats, from 25.4 mm (1 in.) to 300 mm (11.8 in.). Semiconductor fabrication plants are defined according to the dimensions of the wafers produced. Currently the most advanced plants produce 300 mm wafers, however new plants are being developed for wafers up to 450 mm diameter.

[6] The success of the Open Lab Innovation Incubator Center has been recognised through numerous international awards. In 2005 it was named best incubator by the Asian Association of Business Incubation, and in 2006 it received the global-level Randall M. Whaley Incubator of the Year award from the National Business Incubation Association (USA).

engineering and strategy. Many graduates of the college have gone on to become part of the most innovative companies in the LIS, or have founded their own firms.

As noted earlier, the Taiwanese government has always recognised the importance of promoting relations between the various actors in its economic systems. This is reflected in the policies of the ITRI, which from its founding has been highly attentive to favouring inter-firm and firm-research centre relationships. This is demonstrated by the numerous consortium projects established by its research departments.

4.2.1.4 Government Institutions

Government institutions have had a key role in the formation of the Hsinchu LIS. Although the semiconductor sector was already developing autonomously in Taiwan, affirmation of the LIS took place only after the central government, having understood the importance of the Hsinchu scientific and technological park, stimulated all the companies to invest specifically in the semiconductor area.

Another positive aspect of the Taiwan institutions lies in their capacity to enact legislation that favours research and the diffusion of the results. In 1999 the national government proclaimed the Basic Law of Science and Technology, inspired by the American Bayh-Dole Act. This permitted universities and public research centres to patent the innovations developed by their researchers as institutional property, and to derive profits from these through concession of licenses. The statute provided strong stimulus to the individual universities, which oriented their scientific activity towards the creation of innovation suitable for patenting. In addition, the statute encourages the universities to establish technology transfer and "technology trade" centres, for communication and commercialisation of their innovations (Mathews and Hu 2007).

Government has also dedicated numerous forms of financing to both the universities and ITRI, for their work in innovation. Over the 2006–2010 period the Ministry of Economic Affairs, together with the Ministry of Education and the National Science Council, allocated 14.48 million dollars for research in the semiconductor field.

Government efforts have not only been devoted to the research sector. The Ministry of Economic Affairs has also established the Small and Medium Enterprise Administration, which provides consulting activities for small local firms and operates as an incubator for innovative firm creation.

However one of the greatest assets of national government has been its very early realisation of the importance of relationships in the stimulation of innovation. As one example, in order to encourage relations between universities and firms the Ministry of Economic Affairs has instituted the University-Industry Cooperation Steering Committee, which coordinates six Industry-University Cooperative Research Centres, each operating as a platform for the stimulation of encounters, information exchanges and cooperation between members of the research and entrepreneurial spheres.

Table 4.3 Sectoral classification of firms in the Hsinchu cluster

Sector	Number of firms	Employees	Capital (million USD)	Sales (million USD)
Integrated circuits	164	66,467	23,992	22,309
PCs & peripherals	58	14,268	2,267	4,147
Telecommunications	52	7,258	926	1,816
Optoelectronics	61	24,932	4,712	3,927
Precision machinery	21	1,529	185	277
Biotechnology	28	1,023	162	76
Total	384	115,477	32,244	32,552

Data Hsinchu Science and Technology Park

4.2.1.5 The Companies

In spite of the enormous importance assumed by the universities, research centres and institutions, the true heart of the LIS is its companies. There are more than 300, largely specialised in the semiconductor and personal computer sectors, and they compose the principle factor contributing to the area's success (see Table 4.3). There are in fact eight scientific and technological parks in Taiwan, but only the Hsinchu park owes its success to such a myriad of companies.

The structure of the cluster features only a few very large firms, with Acer Inc. being the largest of all, and a myriad of small firms. In the subsections that follow we first examine the roles of the small firms and then the large firms.

Small firms are the core fabric of Taiwan's economy. For this reason, some authors have compared the Taiwan development model to the Italian case (Guerrieri and Pietrobelli 2004). The birth of an entrepreneurial fabric founded on small and medium enterprise can be attributed in part to the local culture, which stimulates the birth of firms that generally follow the model of a family company. Another causal factor is the presence of numerous government policies in support of SMEs, such as the creation of support services and provision of venture capital for financing innovative projects.

For the current study we first examine the role assumed by the SMEs in the Hsinchu LIS and then compare this to the tasks undertaken by the large firms, including Acer. The presence of an entrepreneurial system founded on SMEs has long permitted the Taiwanese economy to blend the characteristics of both flexibility and specialisation. Still, in the passage from an economy based on the manufacturing sector to one based on knowledge (Lin 1997), small size becomes an obstacle. Small firms do not have the necessary resources to invest in research and innovation, activities that are necessary to operate in the knowledge economy. The drive towards the knowledge economy imposes substantial modifications in the local entrepreneurial fabric, encouraging the formation of SME consortia and the birth of large firms (Ernst 2000, 2001).

As noted previously, the Hsinchu LIS was initially composed primarily of small firms that operated as assemblers for large Silicon Valley companies.

Subsequently, with constant government attention to research, many SMEs evolved and began not only to do the assembly of products but also to plan and build more complex components. In this stage, the firms continued to depend on the foreign multinationals. With the growth of some Taiwanese firms, the SMEs then continued as specialised suppliers to the local large firms, gradually reducing their dependence on foreign companies.

Currently there is a separation of roles between the SMEs and the large firms of the Hsinchu LIS. The former are primarily responsible for the production activities while the latter have taken on the function of generating and coordinating relationship among the different actors from the research, institutional and entrepreneurial spheres. At the same time, the large firms also coordinate relations between the system and the external actors, meaning the international consumers and suppliers. This coordinating aspect in the roles of the large firms is examined in the following subsections.

4.2.2 Acer Incorporated

Acer Inc. was originally formed in 1976, under the name Multitech. The company was founded by Stan Shih, his wife Carolyn Yeh and three colleagues known to each other since their university years. In little more than 30 years it became the third producer in the world for PCs, after Hewlett Packard and Dell, and the second for notebooks. In more recent years it improved its position further by purchase of two American giants, Gateway and Packard Bell, which now flank the Acer and eMachines brands in the company's PC sales. Acer's extraordinary development is without doubt to be attributed to the firm's inspired entrepreneur, who overturned the systems in the PC sector and created the Taiwan's first IT company with a global brand.

Acer started with only 11 employees as an assembler for American multinationals, like other Taiwanese companies of the period. However the firm began to rise above the others within just a few years.

In 1981, still under the name Multitech, the company launched the "Microprofessor", a low-cost product in the embryonic age of personal computers, which among other tasks could read files from cassette tapes and visualise them on a screen. The next year the company launched Microprofessor II, with a small keyboard interface on one side of the microprocessor case and a stand-alone screen for file visualisation. In 1983 came Microprofessor III, Multitech's first true personal computer, which followed an IBM design but with components and software completely compatible with Apple models, such that the computer was nicknamed the "Asian clone" of Apple.

The next truly great stride took place in 1986 when Multitech launched its first personal computer based on Intel 386 technology, at the same moment as IBM launched its own product with the same chip. At this time, Multitech changed its name to Acer and began its affirmation as a world brand. Shih hired a former manager from IBM to transform the company into a truly global player. Initially the aggressive growth strategy, based in part on international acquisitions, did not bring the desired results. After some difficult years the company recovered, becoming a world leader in PC sales and the first Taiwanese company with a strong brand image.

Acer's development would not have been the same without the enlightened policies of Stan Shih, who understood how the success of a computer company lies not only in its capacity to generate innovation, but also in marketing. Shih emphasised this concept strongly in his "smiling curve" model. According to this, for a firm to capture the maximum of the value added in the PC manufacturing chain, it has to operate in both the upstream stages of construction and assembly of the different subcomponents and components, and the downhill stages of marketing and distribution. This was the source of Shih's idea to create a company capable of carrying out all the stages of the PC production chain.

Acer's founder also held that the problem of Taiwanese firms did not lie in the lack of innovation investment by local companies, but in the poor image that Taiwanese products had in the global market. In fact in the collective imagination, "made in Taiwan" was a synonym for low price and poor quality. Shih stressed the need to eliminate this effect, which he called the "MIT stereotype", by investing heavily in communications and adopting a price policy similar to those of Japanese computer producers. This allowed Acer to immediately differentiate itself from its local competitors, who generally continued their strategies of cost leadership (Hung and Wittington 2011).

Acer's attention to marketing should not be interpreted as abandonment of the R&D and innovation aspects of the business. Indeed the opposite is true, as demonstrated by the continuous stream of innovations emerging from the company, the number of patents obtained (roughly 1050), and the continuous launch of new products.

One of Acer's objectives is in fact indicated in the motto "Fresh products for everyone", which calls for continuous innovation on the one hand, and on the other the realisation of products that are accessible to all in terms of both price and ease of use.[7]

The specificities of Acer, making it unique in rapidly climbing to world-leading status among otherwise similar firms, lie in three principles of organisation: the "fast food production model", the "global brand, local touch" principle, and the creation of a "client-server organisation".

The first principle refers to the production model adopted by Acer. One of the problems in manufacturing PCs, as with many other high-tech products, lies in their rapid obsolescence. By the time the consumer in a particular country

---

[7] Acer has always considered ease of accessibility as a fundamental aspect of its products. As an indication, the company was the first ever to produce an "instruction manual" for its computers, at the time of Microprofessor I.

demands a good, there is a high risk that the specific corporate product being offered at that location and time is already obsolete. To avoid this problem, Acer divided the fabrication process for its PCs into two stages. The first stage takes place in the home establishment in Taiwan, called the "global kitchen", where the design is defined and the main components are built. The second stage is the final assembly, which takes place at the Acer branches situated near the end market, in what Acer terms "Uniload centers". This organisation ensures that the products entering the commercial marketplace are assembled just a few days prior to delivery, and permits modifications until near the last moment. This guarantees that the consumer always chooses from and receives a current, or "fresh", product.

The "global brand, local touch" principle refers to Acer's internationalisation strategy. Acer has branches throughout the world and its heavy percentage of sales in the United States makes that nation a particularly important end market. If the company had relied only its local market it clearly would have failed to develop or survive. In fact the continuous push for innovation began with the company's initial drive to move into external markets.[8] However PCs are not standardised products, and both their physical and intangible aspects have to be modified continuously in order to adapt to changing local markets. For this reason, when Acer chose the internationalisation route, it did so through creation of joint ventures with local operators in the various markets. These were often distributors, who not only provided rapid market access but also permitted acquisition of the knowledge necessary to render the product as responsive as possible to the needs of the local consumer.

Finally, the "client-server model" refers to company organisation. Acer's organisational structure is quite articulated. At the level below top management are several strategic business units that are responsible for technological development, production of key components and marketing. Alongside these are a series of regional business units, which provide assembly services and are responsible for strategic management in local markets. The unusual feature of this organisational model lies in the autonomy afforded to the individual units and peripheral managers. These have ample decision-making room and they coordinate with the other units in client-supplier types of relationships. The decentralised approach stimulates the birth of ideas and innovations, improves personnel satisfaction, and creates an atmosphere of healthy internal competition. At the same time, the strongly shared company values, highly efficient procedures for inter-company communication, and the presence of charismatic leadership maintain the cohesion of the overall company system.[9]

---

[8] Many studies of the development of multinationals in emerging economies demonstrate similar situations, where the will of the company to assert itself in international markets pushes them to innovate and grow.

[9] Stan Shih provides a personal web site, including recordings from his speeches and writings on his ideas (Chinese only).

4.2.3 Acer and the Development of the Hsinchu LIS

In this section we conduct an analysis of Acer's relationships within the Hsinchu LIS, through classification of the different channels and initiatives utilised by the company to favour diffusion of innovation and encourage the entrepreneurial spirit.

4.2.3.1 The Sub-suppliers Network

The preceding sections have described how the Hsinchu LIS evolved and how small local firms, without the capacity and the economic strength to deal directly with the market, have always operated in a manufacturing role for other companies. Until the 1990s the larger companies were always the Silicon Valley multinationals. Later, as the relations with the American cluster blossomed, the foreign multinationals were gradually substituted by large local companies.

Some small companies now operate on behalf of Acer as assemblers, but many more serve as full manufacturers of technologically advanced components. For some types of production, Acer continues to operate on behalf of the foreign multinationals but subcontracts the production to the smaller local firms. In fact Acer serves as a bridge between the small local and large high-tech multinationals (Ernst 2000, 2001). Many of the high-tech foreign firms prefer to relate with a single large-sized local counterpart rather than many small suppliers.

These types of relationships are governed by two primary means (Guerrieri et al. 2001; Guerrieri and Pietrobelli 2004):

- Contractual agreements—create true inter-sectoral networks that link Acer and its suppliers and assemblers;
- Equity-type relationships—a share of the SME capital is held by the larger firm, which in exchange ensures access to end markets, knowledge transfer, and use of a common brand.

The existence of these links leaves Acer free to maintain a flexible structure. The company, through drawing on its relationships, can adapt continuously to market needs and maintain lower production costs. At the same time, the small firms also achieve advantages. They acquire new knowledge through the construction plans accompanying their incoming orders and develop experience in the production of specific products. The can achieve economies of scale, access credit more easily, and reduce the costs and risks involved in their activities.

However recent years have seen a gradual reduction in the number and intensity of such links in the Hsinchu LIS. Many Taiwanese firms are transferring their production activities to the Suzhou area of China to draw on lower costs of manpower. Acer itself has also transferred many of its sub-supplier links to China and other emerging nations (Mathews 2002; Dedrick and Kraemer 2006).

4.2.3.2 The Spin-Offs

Acer's role in the Hsinchu LIS is not only limited to the construction of the suppliers network with existing firms, but also involves the creation of new firms, which enrich the local system of SMEs.

Over the years, Acer went from having a highly integrated production system to one specialised in specific activities in the value chain. This was done through processes of spin-off of the Acer corporate functions, particularly connected with production, which are now completed by sub-suppliers.

A large part of Acer's spin-offs are thus the result of an explicit process of externalisation from the company, governed by the necessity to separate activities for the sales of Acer-brand computers from the activities of the firm as sub-supplier to other large international companies. In many of these spin-offs, at least in the early stages, Acer maintained a shareholder position (see Table 4.4).

4.2.3.3 Patent Licenses

As noted, Acer invests in innovation on a continuing basis. The company currently holds independent title to over 2,300 international patents. Not all of the patented innovations have been implemented by the company, and instead many were immediately released under license. The Acer managers support licence concession very strongly, thus ensuring that company innovations are diffused within the

Table 4.4 Examples of Acer spin-offs

Company	Company purpose	Acer participation		
		2003 (%)	2006 (%)	2008 (%)
Wistron Corp	Computer manufacturing and assembly for third party companies (OEM)	43.38	14.85	4.56
BenQ	Manufacturing consumer electronic products	15.37	5	None
AMBIT	Communications and computing equipment for third parties (OEM)	14.85	None	None
ALi Corp	Semiconductor R&D	34.97	None	None
Aegis Semiconductor	Semiconductor manufacturing	36	44.03	44.03
Yam Digital	Search engines, digital interactive	5	10.13	n.d.
Legend Design Technology	Manufacturing flash drives and memory cards	10.13	10.13	n.d.
Feiya Technology	Flash drives	1.72	1.72	n.d.
RDC Semiconductor	Semiconductor manufacturing	16.88	5.75	None

Data source Acer 2008 annual report and Osiris Database
n.d. No data available

Hsinchu cluster. This philosophy is best expressed in the words of Stan Shih, the company founder:

> ....There is a Chinese TV- commercial slogan: "Sharing good things with your good friends."......Many global companies enjoy possessing proprietary technology, especially the Japanese companies, as it guarantees no rivals in the market for a huge profit. But Acer does not follow that idea. In the high-tech industry, today's unique technology may become tomorrow's cliché. We might as well share the technology with others when it is still worthwhile.
> The advantage is that, on one hand, we can re-invest the royalty income in other new research efforts to make sure the technology is continuously improved. On the other hand, when making the patents available to others, the product will become more popular and more companies will participate in manufacturing it. We can therefore reduce the cost through maximization of production volume and maintain the company's competitiveness at the same time.
> From another viewpoint, the possession of proprietary technology is dangerous. Many companies hold on to their proprietary technology simply because their operating cost is too high and they believe that this is the only way to generate sufficient profits to cover the cost. This approach in reality hides potential problems and delays the timing for correction as needed. Moreover, the company will surely encounter adverse feelings from others, especially if the company places pressures upon other companies or its customers with its technological advantage. It is impossible for the company not to make any mistakes. Once it does, or its technology is no longer a winner, the resistance from others will become even more forceful.
> As a result, it will bring even greater operating risks to the company.[10]

As Acer's founder emphasises, the concession of patent licenses to other companies in the system in the manner of friendship permits diffusion of the knowledge and obtains financial resources that ensure further investment in innovation.

4.2.3.4 Labour Mobility

Acer also contributes to growth of the system through mobility of its employees. Studies by Lee and Saxenian (2008) reveal that employee mobility is very high among employees of Taiwanese firms. On average, a worker remains with one company for a period of about 3 years before transferring to a larger firm or creating their own entrepreneurial activity.[11]

Numerous former Acer employees are now employed in other firms in the LIS. Workers are not only free to move, in some senses this kind of mobility is actually encouraged by the company. As Stan Shih recounts, Acer was born with the concept of serving as a training centre, and over the years it has maintained this philosophy to some extent:

> ......when founding Acer 28 years ago, I made up mind to be the master without secrets. For the last 28 years along Acer's successful track, we've always tried to share all our

[10] Shih (2006) *Me Too Is Not My Style*. Acer Publications, Taiwan.

[11] The characteristic of worker mobility is what permitted the initial formation of the Hsinchu cluster. Without the fact of the engineers employed at Silicon Valley returning to Taiwan to form their own companies, the national semiconductor industry would never have been born.

> experiences, good or bad. In today's Taiwanese high-tech industry, Acer is referred to by many people as "Acer University," since a lot of the key persons in the brand name companies, IT manufacturing companies, and international marketing companies of the industry were either employed or trained by Acer. And fortunately, I am glad to say that Acer is still one of the most successful companies.
> Acer's principle management philosophy is empowerment. We believe in the truth of empowering people. We believe in decentralized management, which means we ask our employees to be the boss of their small division. We asked our managers in affiliated companies to take the driver's seat. We try to share the experience as well as the profit. We are also willing to pay course fees to develop talent. All of the above is important philosophy in the success of Acer. In Acer, we empower people not just for the sake of Acer. We are happy to see all employees gear up to their full potential to become chairmen or CEOs in Acer. And we are just as happy to see them gear up to their full potential and be successful outside Acer. [12]

The link between Acer and a firm that takes in one of its workers permits knowledge transfer between the two actors in the LIS, without monetary transactions. Such links provide the basis for creation of a shared culture within the innovation system, since Acer's values and knowledge are transferred to the other company via the worker.

4.2.3.5 The Managers' Role

As a group, Acer's managers represent an enormous body of knowledge. Sharing this knowledge with the smaller firms in the system is one of the larger company's fundamental contributions to the Hsinchu LIS.

The affective links deriving from shared experience are an important component of the life of the Taiwanese manager, and represent a factor that boosts dialogue and information sharing. The Taiwanese business and economic systems are fed by informal relations between the various actors, a characteristic that may also be seen in other nations with a Confucian heritage. Informal groups often develop between managers belonging to different companies, who then transfer information and share their experiences.

One of the important occasions for sharing is at the annual alumni meeting organised by the National Chiao Tung University. The university graduates include the founder of Acer and the top managers of many other Hsinchu LIS companies. These meetings maintain and create new relationships between managers, which solidify as links between their respective companies.

Information transfer between one firm and another can also take place through the participation of the Acer board members in the boards of directors of other companies in the system. The individual managers function as channels for knowledge transfer between the companies where they are directors. Such channels are particularly important when these cases of board interlock are the only link between Acer and the other company.

[12] Excerpt from "Building up trust in business", keynote address by Stan Shih to the World Young Leaders Forum.

Table 4.5 Examples of interlock between Acer and other companies in the LIS

Role in Acer	Name	Role in other company	Company activity
CEO	J.T. Wang	Chairman of HiTRUS Inc.	Security services for on-line payments
Board of directors	Stan Shih	Director of Dragon Investment Co.	Investment fund
		Director of Qisda Corp.	LCD monitor manufacturing
		Director of Wistron Corp.	PC manufacturing
		Director of iD SoftCapital	Venture capital, business incubator
Manager	G. Huang	Director of Apacer Technology Inc.	RAM and Flash memory manufacturing
Board of directors	Philip Peng	Manager of Acer Laboratories Inc.	Software R&D
		Manager of Wistron Corp.	PC manufacturing
		Manager of Apacer Technology Inc.	RAM and Flash memory manufacturing
		Director of iD SoftCapital Inc.	Venture capital, business incubator
		Manager of Dragon Investment Co., Ltd.	Investment fund
Manager	Carolyn Yeh	Director of iD SoftCapital	Venture capital, business incubator

Sources Acer annual report, Osiris Database, corporate websites

An analysis of board interlock, presented in Table 4.5, shows the roles of Acer managers in other companies. The Acer directors are particularly represented in the boards of firms that were created by spin-off from Acer, even if Acer is not a shareholder in the new company, as for example in the case of ALi Corp.

In some cases the Acer managers launch new entrepreneurial initiatives, while still remaining as employees of the larger multinational.[13] For example Stan Shih and other senior managers of the group invested their own assets to form iD SoftCapital Inc., thus taking a personal leadership role in renewal and development processes for the Hsinchu LIS.

The aim of iD SoftCapital Inc. is to support innovation and development in the local system through consulting services. The specific services offered are:

- development of venture capital financing, through investment from the sister companies, iD Techventure and iD Venture America;
- attainment of support services for the birth of new companies, through a further sister company, iD Innovation;
- development of financing for existing companies to re-engineer and promote their brands, through iD Reengineering and iD Branding.

13 Since the managers are still Acer employees, we do not considered these new initiatives as true spin-offs.

iD SoftCapital plays an important role in supporting the evolution of the cluster, by ensuring that new companies can come into being, as well as by supporting the existing ones.

4.2.3.6 Acer, the Universities and Other Education

It is difficult to identify the potential leadership role of Acer in regards to the two universities of the Hsinchu LIS. However since there is a tradition in Taiwan of alumni donating funds to their home universities, and since the universities in turn depend on these funds for important budgets, we can imagine that Stan Shih contributes substantial sums to his alma mater, the National Chiao Tung University.

It is also interesting to observe the initiatives adopted by Acer for diffusion of a culture of innovation among young students in the area. When these eventually enter employment they will work for a variety of firms throughout the LIS, thus spreading the benefits of the Acer "investments". Two of the specific initiatives to note here are the annual Long Term Smile Contest and the Digital Creation Award, both directed at school-age children.

Acer also participates actively in continuous development of the new managerial class, through its Aspire Academy. This is a centre for managerial education financed by the Acer Foundation, open to all Taiwanese graduates, offering training programs and courses in management, marketing, and strategy in science and technology. The academy is the first step in plans for a broader project to develop Aspire Park, a true science and technology park just outside the Hsinchu area. One of the essential elements of the new park is a business incubator, with responsibilities for financing the birth of new innovative companies.

Aspire Academy permits Acer to transfer part of its knowledge heritage to the student participants, who in turn transfer the knowledge to their various employer companies within the Hsinchu LIS. Acer's training activities thus also create and strengthen the links between the larger firm and the other actors.

4.3 Conclusions

In conclusion, in this chapter we have seen how Acer contributes to the development of a Taiwanese LIS through favouring knowledge exchanges and creating a culture of innovative entrepreneurialism. The major mechanisms used by this large firm to favour the birth of new knowledge and the growth of spillovers are:

- establishing sub-supplier relationships with other firms in the cluster, which include transfer of knowledge through the process of providing instructions for building components;
- maintaining close relationships with budded firms;
- diffusing new innovations through concession of patent licenses;
- permitting labour mobility, and in some cases favouring it, and in this way transferring the Acer culture and knowledge to the smaller firms.

Through these initiatives, Acer favours the birth of innovative ideas, but above all it creates the entrepreneurial atmosphere that serves as the glue for long-term relationships between the system members, thus favouring its entire function.

References

Agrawal A, Cockburn I (2003) The anchor tenant hypothesis: exploring the role of large local R&D-intensive firms in regional innovation systems. Int J Ind Organ 21:1227–1253

Albino V, Garavelli AC, Schiuma G (1999) Knowledge transfer and inter-firm relationships in industrial districts: the role of the leader firm. Technovation 19:53–63

Amin A, Cohendet P (2000) Organisational learning and governance through embedded practices. J Manage Gov 4(1–2):93–116

Asheim BT (1994) Industrial districts, inter-firm cooperation and endogenous technological development: the experience of developed countries. In: UNCTAD (ed) Technological dynamism in industrial districts: an alternative approach to industrialization in developing countries? United Nations conference on trade and development, New York and Geneva, pp 91–142

Audretsch D, Feldman MP (1996) R&D spillovers and the geography of innovation and production. Am Econ Rev 86(3):630–640

Baptista R (2000) Do innovation diffuse faster within geographical clusters? Int J Ind Organ 18:515–535

Becattini G (1989) Sectors and/or districts: some remarks on the conceptual foundations of industrial economics. In: Goodman E, Bamford J (eds) Small firms and industrial districts in Italy. Routledge, London, pp 123–135

Boari C, Lipparini A (1999) Networks within industrial districts: organising knowledge creation and transfer by means of moderate hierarchies. J Manage Gov 3:339–360

Breshanan T, Gambardella A, Saxenian AL (2001) 'Old economy' inputs for 'new economy' outcomes: cluster formation in the new silicon valleys. Ind Corp Change 10(4):835–860

Camagni R (1991) Local milieu, uncertainty and innovation networks: towards a new dynamic theory of economic space. In: Camagni R (ed) Innovation networks: spatial perspectives. Belhaven Press, London, pp 121–142

Camuffo A, Grandinetti R (2011) Italian industrial districts as cognitive systems: are they still reproducible?. Entrepreneurship Reg Dev 23:815–852

Camuffo A, Romano P, Vinelli A (2001) Back to the future: Benetton transforms its global network. MIT Sloan Manag Rev 43(1):46–52

Cantwell J, Iammarino S (2003) Multinational corporations and European regional systems of innovation. Routledge, London, New York

Carbonara N (2004) Innovation processes within geographical clusters: a cognitive approach. Technovation 24:17–28

Chang TC (2009) Systems of innovation, spatial knowledge links and the firm's innovation performance: towards a national-global complementarity view. Reg Stud 43(9):1199–1223

Chen TJ (2008) The emergence of Hsinchu science park as an IT cluster. In: Yusuf S, Nabeshima K, Yamashita S (eds) Growing industrial clusters in Asia: serendipity and science. World Bank, Washington DC

Chen S, Choi CJ (2004) Creating a knowledge-based city: the example of Hsinchu science park. J Knowl Manag 8(5):73–82

Chyi YL, Lai YM, Liu WH (2012) Knowledge spillovers and firm performance in the high-technology industrial cluster. Res Policy 41(3):556–564

Dedrick J, Kraemer KL (2006) Is production pulling knowledge work to China? a study of the notebook PC industry. Computer 39(7):36–42

Dei Ottati G (2009) An industrial district facing the challenges of globalization: Prato today. Eur Plan Stud 17:1817–1835

Dhanaraj C, Parkhe A (2006) Orchestrating innovation networks. Acad Manag Rev 31(3):659–669
Edquist C (2005) Systems of innovation: perspectives and challenges. In: Fagerberg J, Mowery D, Nelson RR (eds) The oxford handbook of innovation. Oxford University, Norfolk
Ernst D (2000) Inter-organizational knowledge outsourcing: what permits small Taiwanese firms to compete in the computer industry? Asia Pac J Manag 17:221–255
Ernst D (2001) Small firms competing in globalized high-tech industries: the coevolution of domestic and international knowledge linkages, in Taiwan's computer industry. In: Guerrieri P, Iammarino S, Pietrobelli C (eds) The global challenge to industrial districts: SMEs in Italy and Taiwan. Edward Elgar, Cheltenham, pp 95–130
Etzkowitz H, Leydesdorff L (2000) The dynamics of innovation: from national systems and 'mode 2' to triple helix of university-industry-government relations. Res Policy 29(2):109–123
Feldman MP (2003) The locational dynamics of the US biotech industry: knowledge externalities and the anchor hypothesis. Ind Innov 10(3):311–329
Feldman M, Lowe N (2008) Consensus from controversy: Cambridge's biosafety ordinance and the anchoring of the biotech industry. Eur Plan Stud 16:395–410
Giuliani E, Bell M (2005) The micro-determinants of meso-level learning and innovation: evidence from a Chilean wine cluster. Res policy, 34(1):47–68
Glasmeier A (1991) Technological discontinuities and flexible production networks: the case of Switzerland and the world watch industry. Res Policy 20:469–485
Grabher G (1993) The weakness of strong ties: the lock-in of regional development in the Ruhr area. In: Grabher G (ed) The embedded firm. Routledge, London, pp 1–32
Guerrieri P, Pietrobelli C (2004) Industrial districts' evolution and technological regimes: Italy and Taiwan. Technovation 24:899–914
Guerrieri P, Iammarino S, Pietrobelli C (2001) The global challenge to industrial districts: SMEs in Italy and Taiwan. Edward Elgar, Cheltenham
Hobday M (1995) East Asian latecomer firms: learning the technology of electronics. World Dev 23(7):1171–1193
Hu T, Lin C, Chang S (2005) Role of interaction between technological communities and industrial clustering in innovative activity: the case of Hsinchu District, Taiwan. Urban Stud 42(7):1139–1160
Hung SW, Tang RH (2008) Factors affecting the choice of technology acquisition mode: an empirical analysis of the electronic firms of Japan, Korea and Taiwan. Technovation 28:551–563
Hung SC, Whittington R (2011) Agency in national innovation systems: institutional entrepreneurship and the professionalization of Taiwanese IT. Res Policy 40(4):526–538
Ku YL, Liau SJ, Hsing WC (2005) The high-tech milieu and innovation-oriented development. Technovation 25:145–153
Lazerson M, Lorenzoni G (1999) The firms that feed industrial districts: a return to the Italian source. Ind Corp Change 82(2):235–266
Lee CH, Saxenian A (2008) Coevolution and coordination: a systemic analysis of the Taiwanese information technology industry. J Econ Geogr 8:157–180
Lin CY (1997) Technopolis development: an assessment of the Hsinchu experience. Int Plann Stud 2(2):257–273
Link A, Scott JT, Siegel DS (2003) The economics of intellectual property at universities an overview of the special issue. Int J Ind Organ 21:1217–1225
Lissoni F (2001) Knowledge codification and the geography of innovation: the case of Brescia's mechanical cluster. Res Policy 30:1479–1500
Lorenzoni G, Baden-Fuller C (1995) Creating a strategic center to manage a web of partners. Calif Manag Rev 37(3):146–163
Maskell P, Malmberg A (1999) The Competitiveness of Firms and Regions 'Ubiquitification' and the Importance of Localized Learning. Eur Urban Reg Stud 6(1):9–25

Mathews JA (2001) The origins and dynamics of Taiwan's R&D consortia. Res Policy 13(15):1–20
Mathews JA (2002) Dragon multinational: a new model for global growth. Oxford University Press, New York
Mathews JA, Hu MC (2007) Enhancing the role of universities in building national innovative capacity in Asia: the case of Taiwan. World Dev 35(6):1005–1020
Morrison A (2008) Gatekeepers of knowledge within industrial districts: who they are, how they interact. Reg Stud 42(6):817–835
Munari F, Sobrero M, Malipiero A (2011) Absorptive capacity and localized spillover: focal firms as technological gatekepeers in industrial districts. Ind Corp Change 8:1–34
Niosi J, Zhegu M (2005) Aircraft systems of innovation. In: Niosi J (ed) Canada's regional innovation systems. McGill-Queen's University Press, Montreal, Kingston, pp 61–86
Niosi J, Zhegu M (2010) Anchor tenants and regional innovation systems: the aircraft industry. Int J Technol Manage 50(3–4):263–284
Parilli MD (2009) Collective efficiency, policy inducement and social embeddedness: drivers for the development of industrial districts. Entrepreneurship Reg Dev 21:1–24
Parmentola A (2009) Il cluster tecnologico di Hsinchu e il ruolo dell'Acer. In: Simoni M, Schiavone F (eds) Il Ruolo Della Grande Impresa Nello Sviluppo Dei Poli Innovativi. Mcgraw-Hill, Milan
Salancik GR, Pfeffer J (1978) A social information processing approach to job attitudes and task design. Adm Sci Q 23:224–253
Saxenian A (1991) The origins and dynamics of production networks in Silicon Valley. Res Policy 20(5):423–438
Saxenian A, Hsu JY (2001) The silicon Valley-Hsinchu connection: technical communities and industrial upgrading. Ind Corp Change 10(4):893–920
Shih S (2006) Me too is not my style. Acer Publications, Taiwan
Simmie J (2002) Knowledge spillovers and reasons for the concentration of innovative SMEs. Urban Stud 39(5–6):885–902
Utterback J (1994) Mastering the dynamics of innovation: how companies can seize opportunities in the face of technological change. Harvard Business School, Cambridge
Van Winden W, Van Den Berg L, an Der Meer A (2004) The development of ICT clusters in European cities: towards a typology. Int J Technol Manag 28:356–387
Wolfe DA, Gertler MS (2004) Clusters from the inside and out: local dynamics and global linkages. Urb Stud 41(5–6):1071–1093
Wong PK (1999) National innovation systems for rapid technological catch-up: an analytical framework and a comparative analysis of Korea, Taiwan and Singapore. Presented at the DRUID summer conference on national innovation systems, industrial dynamics and innovation policy, Rebild, Denmark, 9–12 June 1999
Yang DYR, Hsu JY, Ching CH (2009) Revisiting the silicon island? The geographically varied 'strategic coupling' in the development of high-technology parks in Taiwan. Reg Stud 43(3):369–384

Chapter 5
The University-Driven LISs

Abstract Theoretical conceptions of the role of universities in local innovation systems have evolved over the past 20 years, from an initial approach that stressed the importance of knowledge spillovers from university educational and research activities into their regional knowledge spaces, towards a new emphasis on the "third role" of universities, as animators of regional economic and social development. However the typical conception is that universities and research centres in developing countries take little part in stimulating the formation of LISs, with this mission instead being substituted by the state or large companies. This chapter analyses the role and the mechanisms through which universities can contribute to creation of LISs, first from a theoretical point of view. The theoretical arguments are then supported by an analysis of the role of the National University of Singapore, in driving the development of the "Biopolis" biotech LIS.

Keywords LIS · Universities · Third mission · Singapore · Biopolis cluster · NUS

5.1 The Role of Universities in Local Innovation Systems

Universities have long been recognised as providers of basic scientific knowledge for industrial innovation, through their actions in research and related activities. Such benefits were understood as particularly accruing to the agricultural and manufacturing sectors (Guston 2000; Smith 1990; Hart 1988).

Neoclassical economic theory explained the productive performance and competitive advantage of firms largely in terms of relative resource endowments (Hall 1994). The role of knowledge and of institutions involved in the creation of knowledge was seen as exogenous, though not unimportant, to the production system (Freeman 1995). The emergence of the national systems of innovation approach (Freeman 1991; Lundvall 1992) shifted the conceptualisation of the role of universities in economic production, bringing them "inside the tent", among the other actors with a direct role in shaping regional innovation systems (Cassia et al. 2008).

M. Ferretti and A. Parmentola, *The Creation of Local Innovation Systems in Emerging Countries*, SpringerBriefs in Regional Science,
DOI 10.1007/978-3-319-10440-9_5

There are two dominant approaches to perceiving this role: the triple helix model of university, industry and government relations, and the concept of the "engaged university", expressed in a second body of literature. The two concepts overlap, but also have important differences in emphasis. Both bodies of thought stress that universities are increasingly linked to place but they offer different analyses of the driving forces in the relationship. Further, there are significant differences in their assumptions regarding institutional norms and behaviours.

The triple helix model (Etzkowitz and Leydesdorff 1997) brought attention to the role of universities in regional economies, anticipating the multiplying incidence of hybrid university-industry-government relationships, as seen in recent resource and capital development projects. Particularly interesting among these projects are those involving property development for science parks and incubator facilities, to favour new firm formation (Etzkowitz 2002: 14).

The triple helix model offers a conceptualisation of innovation as occurring through non-linear, interactive processes, in a "recursive overlap of interactions and negotiations among universities, industry and government" (Etzkowitz and Leydesdorff 1997). One of the key insights of the model is that of the hybrid, repetitively growing, inter-institutional nature of relations among the "three helices". Instead of the traditional situation in which the institutional spheres of the state, university and industry were separate entities that interacted across strongly defended boundaries, the individuals and organisations within these helices are seen as taking different and mixed roles (Etzkowitz and Leydesdorff 1999: 113; Etzkowitz and Leydesdorff 1997; Sutz 1997).

Like the triple-helix model, the literature on the "engaged university" (OECD 1999; Holland 2001; Chatterton and Goddard 2000; OECD 2007; Uyarra 2010) also focuses on the third role of universities in regional development. However it differs in emphasising the "adaptive responses" open to universities as they embed a stronger regional focus in their teaching and research missions. The approach does not eschew the development of hybrid, boundary-spanning mechanisms for external engagement, but it inserts this into a broader, developmental approach that considers a range of mechanisms through which universities can engage with their regions (Power and Malmberg 2008). Universities, through their resource base of people, skills and knowledge, increasingly play a diversity of roles in regional networking and institutional capacity building. Staff, acting in both formal and informal capacities, act as "regional animators" (Chatterton and Goddard 2000: 481) through their insertion in outside bodies ranging from school boards and local authorities to cultural organisations and development agencies. Universities thereby make an important indirect contribution to the social and cultural basis of effective regional governance. In general, the developmental role of universities in regional economic and social development centres on the intersection of learning economies and the regionalisation of production and regulation.

However, the literature on the engaged university appears to downplay the empirical fact of differences in the missions of relevant institutions, and can be argued to oversimplify the capacity and willingness of universities to adapt their functions in response to external signals.

Both of these bodies of literature tend to distinguish the "generative" and "developmental" roles of universities in local innovation systems (Cooke 2002; Niosi and Bas 2001; Lundvall and Johnson 1994).

Universities and public research organisations have been given a much more prominent role in recent models of knowledge production (Charles 2006). Gunasekara (2006: 143) identifies two models of university involvement in regional development. First, the generative role serves regional needs directly, by providing boundary-spanning activities like incubators and science parks. Second, a broader developmental role is filled by adjusting research and teaching activities to regional needs.

In emerging nations, universities now play key roles in LISs. First, they can enhance the regional knowledge base through their international academic networks, serving as gateways for local businesses to reach external knowledge (Altbach 1998: 179; Fritsch and Schwirten 1998). Second, they can adapt knowledge from extra-regional sources to produce new forms that are more appropriate for the local innovation system. In doing so, they reduce entry costs for new technologies and open windows of opportunity for catch-up processes (Perez and Soete 1988: 476). Third, within the centralised character typical of innovation systems in developing countries, public universities usually enjoy greater autonomy than do other regional actors, maintaining substantial levels of control over their financial, personnel, and academic affairs. Universities that are fully or partially incorporated tend to establish even greater levels of independence, and are capable of responding still more efficiently to regional needs.

Bernardes and Albuquerque (2003: 868–870) conclude that linkages and feedback mechanisms among universities and industry have to be established in coevolution, during the catching-up processes. In the worst case, purely education-led growth will encounter a dead end if the university outputs have not been correlated to needs in the productive sector.

In the nascent innovation systems of developing countries, university-industry linkages differ markedly from those in the western experience. Emphasis is placed on human capital development with strong feedback from the private sector, the adaptation of innovations from more advanced countries, and the diffusion of appropriate technologies to local companies (Schiller 2006a, b). The bulk of new knowledge and technologies is traditionally acquired from extra-regional sources, for example via technology licensing and foreign-owned affiliates (Liefner et al. 2006). Universities are the only endogenous knowledge source in many developing-nation LISs and thus take wider responsibility in the whole process of economic development. Local SMEs are often almost completely lacking in technological capacities, and are in need of basic technical education and services, generally without advanced research.

Since the academic capabilities of universities in developing countries are expected to be low at the beginning of the catch-up process, knowledge transfers in projects with large or foreign-owned companies might in fact run from industry to university (Schiller and Liefner 2007).

In addition to the institutional barriers that limit university-industry linkages everywhere in the world, developing countries demonstrate several further restrictions that affect the efficiency of knowledge transfers (World Bank 2000). In such nations, faculty are still tied to the dominant duty of teaching, and may not be sufficiently qualified to provide students with up-to-date skills. Advanced research is hindered by a lack of equipment and funding, and academics' incomes are too low to attract top personnel. Sideline activities that generate additional income may in fact detract from academic duties. In addition, bureaucratic "red tape" at public universities can be burdensome.

5.1.1 Mechanisms for University Guidance of LIS Development

A university that embraces its role and adopts the mission of contributing to regional or national development is referred to as an "entrepreneurial university". According to Etzkowitz et al. (2000), an entrepreneurial university is one that undertakes activities "with the objective of improving regional or national economic performance as well as the university's financial advantage and that of its faculty."

Prior to the triple helix model, the university was conceived as having the traditional two-fold mission of transferring knowledge through education, while also advancing knowledge through basic research. Adding to this dual role, the entrepreneurial university adopts the third mission of contributing to economic development.

The potential activities for achieving this third aim cover a broad and continuous spectrum, ranging from those originating in the traditional university roles to those that express the institution's new entrepreneurial nature, as suggested in Fig. 5.1.

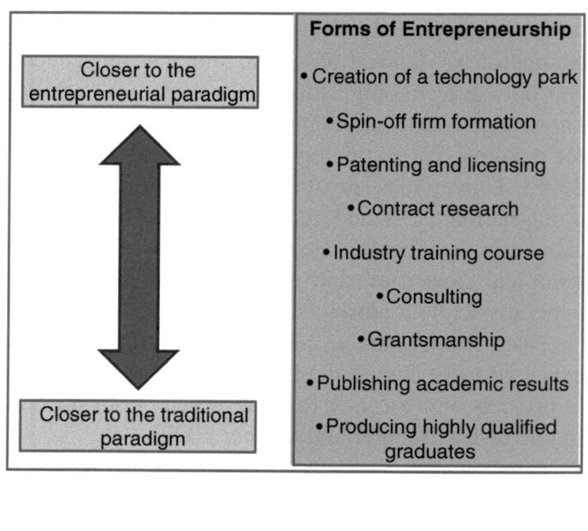

Fig. 5.1 Spectrum of activities for the university "third role" from Philpott et al. (2011)

Considering the possible range of such activities, we distinguish three types of mechanisms that the university can apply to guide local development: (i) start-up development, (ii) university-industry collaborations, (iii) formation and research.

5.1.1.1 Start-up Development

This category includes all initiatives where the university is an actor in creating the necessary conditions for support or guidance in the birth of new innovative firms in the local context. Such activities could take place through participation in science and technology parks, which create a formal site where firms (normally high-tech firms) can locate and interact with the university. These parks provide a seed bed for the development of multiple new ventures and can thus contribute to regional cluster development and employment. The infrastructure in turn contributes to the research capability of the university by attracting highly skilled individuals and technological resources to the region.

Another instrument for universities to promote entrepreneurialism is the creation of academic spin-offs, born for the commercial exploitation of research results. The creation of these new entrepreneurial ventures transfers technology from the lab to market, thus exploiting intellectual property and generating regional employment. Finally, the university can award licenses for use of its patented innovations. The original act of patenting the innovation documents the knowledge contribution of the university and allows the controlled transfer of the intellectual property to selected industrial partners, who can exploit the innovation for competitive advantage and wealth generation. Such licensing can both support existing firms and favour the creation of new entrepreneurial activities (Chang et al. 2009; O'Shea et al. 2008; Wright et al. 2006; Murray 2004; Shane 2004a, b; Di Gregorio and Shane 2003).

5.1.1.2 University-Industry Collaboration

The university can collaborate with firms through a number of mechanisms, for development of various research activities.

Contract research facilitates company activities by solving practical problems that would otherwise deter business performance. Engaging in contract research also contributes to stronger social relations between university and industry that can lead to deeper research interactions in the future.

The university can institute training courses for skills improvement in the national or regional workforce in emerging areas of industrial technology and practice. This ensures that regional industry continually increases its internal skills base and maintains or advances its competitive position.

The university can also provide consultancy services, including the provision of personalised advice and mentoring for improvement in enterprise performance. All of these mechanisms develop linkages between university and industry that can be further exploited in the future. The university-firm relationship is in fact typically bidirectional, since the university will gain advantages on its own side of the cooperative activities. The potential benefits include company investment in university activities through mechanisms such as grants. Such investment enhances the reputation of the university, which attracts further industry to the region, and also tends to lead to development of "hard" academic entrepreneurship by individual faculty members (Powers 2004; Di Gregorio and Shane 2003; Van Looy et al. 2004).

5.1.1.3 Education and Research

Obviously universities will continue to contribute to territorial development through their traditional activities of education and diffusion of research results. The development of specialised personnel capable of meeting current and future demand ensures that national industry can meet its staffing needs, and that it has the absorptive capacity to engage with university as part of the triple helix model.

As with the case of corporate funding for university research, the overall production and diffusion of research results enhances the reputation of the institution, attracting further industry and leading to harder forms of academic entrepreneurship (Powers 2004; Di Gregorio and Shane 2003; Van Looy et al. 2004).

5.2 The Biopolis LIS of Singapore

Singapore's continued competitiveness in the global knowledge economy is based on an ongoing series of government-driven upgrading and renewal programmes that steer institutions and the workforce towards swift adaptation in response to technical and economic trends (Yue and Lim 2002). In 2010, the Singapore Economic Strategies Committee issued a report presenting the key recommendation that over the coming decade, national growth should be productivity-driven, and that such growth should in turn be centred on the fostering of innovation.

Over the years, Singapore has indeed worked towards a comprehensive R&D ecosystem, comprising public sector research bodies, academic research institutes and corporate R&D laboratories. However R&D is not an end in itself. To serve as a driver of economic growth, enterprises must be able to create value from R&D investments and the resulting intellectual properties. Large domestic and foreign corporations, which benefit from strong state backing, have traditionally carried out the bulk of Singapore's innovation and commercialisation activities. For example for many years, the Economic Development Board has offered incentives to large foreign corporations to either relocate or build their regional R&D centres in Singapore. As a result, foreign corporations and large local enterprises accounted for 85.7 % of industry R&D expenditure from 2002 to 2010. Any innovation capabilities in the SME sector have largely developed through cooperation in testing for commercialisation, or from other spillover mechanisms.

Beginning with the 2010 decade, the national government has put increasing emphasis on broadening the base of innovative activities and bringing in local enterprises, especially SMEs (OECD 2013). In Singapore, such policies have the possibility of achieving relatively immediate effect in local development, given the nation's status as a city state. In this case, the national and regional innovation systems effectively coincide.

State promotion of the biomedical industry is just one aspect in the restructuring of the economy, from the former services and manufacturing basis towards a truly knowledge-based economy (Wong 2006a, b). The biomedical sector is part of the planned life sciences "pillar" of the national economy, flanked in turn by the further three pillars of the electronics, chemicals and engineering areas. The life sciences pillar as a whole consists of the core areas of pharmaceuticals, biotechnology, medical technology and healthcare services.

Singapore's State biomedical initiatives began in the 1980s, but it was only in 2000 that the biomedical industry was targeted as a core manufacturing area for major investment of national resources (Lee and Tee 2009). The State established three key agencies, charged with the long-term responsibility of creating a distinctive environment for the biomedical industry, from an international point of view.

The most important of these agencies is A*STAR, charged with the "creation and utilisation of intellectual capital, and the training of research manpower in the transition to a knowledge-based economy" (MTI 2006: 9). The agency was formed in 2001, based in part on the former National Science and Technology Board. It is described as "a luminous Constellation, charting the course of Singapore's Science and Technology progress, comprising the Biomedical Research Council (BMRC), the Science and Engineering Research Council (SERC), Exploit Technologies Pte. Ltd. and the A*STAR Graduate Academy" (Wong 2011).

The second agency is the Biomedical Science Group (BMSG), established under the Economic Development Board (EDB) of Singapore. BMSG is responsible for providing support to biomedical firms that set up R&D and manufacturing facilities, headquarters, or other high-value operations in Singapore. The third agency, under the EDB's investment arm, is Bio*One Capital, which manages investment funds for strategic biomedical technology and start-ups. These three government arms thus have different and complementary roles in the establishment of a biomedical science hub in Singapore (Lee and Tee 2009).

In the context of the industry-wide challenges in international pharmaceutical and biotech markets, Singapore has established itself as a highly conducive place to undertake research. The state has cultivated a network of government agencies, hospitals, manufacturing facilities, universities, R&D infrastructure, contract research agencies, and has provided for access to different workforce talents. The overall strategy has led to the development of a highly innovative culture. This, combined with significant government funding (S$16 billion to date, and S$3.7 billion over 2011–2016) and the promotion of public-private partnerships, has stimulated rapid expansion in Singapore's biotech industry. With this expansion has come the creation of many new jobs across the sector. The increase in employment opportunities provides prospects for local pharmaceutical and biotech professionals, as well as for foreign professionals who wish to enter further in the global market by taking up positions in Singapore.

The various policy aspects for the biotech sector came together in concrete form in 2003, with the creation of Biopolis Shared Facilities/Biomedical Sciences Institutes. "Biopolis" is a high-profile technology park, a physical space that brings together key Singapore biomedical research institutes, national governance bodies, and global and local biotech and pharmaceutical companies. The facility is intended both for internal economic development and as a platform to project Singapore's expertise into the region and the world. The concept of the technology park is to project Singapore as the "Biopolis of Asia", in fact to become the premier life sciences hub in a region where several larger States (China, India, Taiwan, South Korea) are also scaling up public investment in regenerative medicine and genomics (Wong 2006a, b; Salter 2007).

Biopolis has three principal objectives. The first is to serve as a focal point for scientific talent. This means attracting top personnel to carry out world-class research, and providing a fertile training ground for undergraduate and graduate students. Biopolis' role as a talent magnet is considered as the most important of all its contributions, and most crucial to growth of the biomedical industry. The second objective is to integrate and synergise the capabilities and resources of A*STAR's research institutes and to encourage cross-disciplinary research. The third is to bridge private and public sector research work by creating an environment that fosters exchange of ideas and close collaboration. The heightened interchanges between industry and research centre scientists will accelerate the translation of new discoveries to marketable products.

In joining with the National University Hospital and the National University of Singapore, Biopolis becomes a third connecting link in the biomedical knowledge channel, between academic biology and clinical applications.

Biopolis is the biomedical component of a larger technopole called "one-north", a 200 ha cluster of new-economy ventures, including local and large multinational ICT and media companies. The name *one-north* refers to Singapore's location slightly north of the equator. The urban design of the technopole features award-winning architectural structures set in a tropical ambience. *One-north* as a whole is a space devoted to creativity, flexibility and intellectual play (Wong and Bunnell 2006). The Biopolis laboratories are housed in buildings with names such as Proteos, Matrix, Centrius, Chromos, Nanos, Helios, Genome, Immunos, Neuros, Synapse and Amnios. The facilities have dedicated spaces for commercial researchers and house five of Singapore's seven biomedical institutes: the BioInformatics Institute, Bioprocessing Technology Institute, Genome Institute of Singapore, Institute of Bioengineering and Nanotechnology and the Institute of Molecular and Cell Biology.

The Singapore Tissue Network, a central tissue banking facility is also located here, along with the newly created Singapore stem cell bank. The global stem cell company ES Cell International and the multinational pharmaceutical company Novartis also have their laboratories and offices here.

Today Singapore has about 4,300 researchers and scientists engaged in in biomedical research. Companies such as Abbott, GSK, Pharmalogicals, Novartis, Merlion, Merck (MSD), S*Bio, Takeda, Inviragen and Cell Research Corp have all localised activities in Biopolis. However the technological park is not limited to pharmaceuticals and biotech. For example there are also a number of important medical device companies, including Medtronics, the world's fourth largest in the sector. The companies noted here, together with the many others active in Biopolis, produce annual outputs of over S$21 in medicines, nutritional products and medical devices for the global market. Among the smaller companies are Illumina, Scigen, Proligo Singapore, Chakra Biotech Pte, Invida Pharmaceuticals, Veredus Laboratories and Innogene Kalbiotech.

Companies like Cytos Biotechnology, bioMérieux, Humalys and Siena Biotech are all active in research collaborations in Biopolis. The combination of big pharmaceutical and biotechnology companies has led to creation of large numbers of life-science jobs, interesting both to the Singaporean and international employment markets. Current growth areas for investment in Singapore include biologics and rare and tropical disease medicine.

The initiatives by the Singapore government have clearly succeeded in attracting international firms to the scientific and technological park, but the National University of Singapore (NUS) has also played an essential role in transforming Biopolis into a true LIS, favouring the birth of local companies and stimulating cooperation between the research and entrepreneurial spheres.

5.3 The National University of Singapore in the Research Context

Established in 1905, the NUS is the oldest and largest public university in Singapore, with a total student enrolment of over 28,000, three-quarters of which are undergraduates. NUS is also the nation's most comprehensive university, followed by the other two public universities, Nanyang Technological University and Singapore Management University, both of which are younger than the NUS. Historically, government policies have emphasised public education. As a result there are no large private Singaporean universities, although there are many diploma-level private colleges and distance-learning programs. However since the late 1990s, the government had encouraged leading overseas universities to establish branch campuses and other operational presences in Singapore.

Until 1991 public R&D was concentrated in the higher education sector, particularly in the NUS, as well as in a small number of government agencies (Wong and Singh 2005). NUS has historically played a key role in Singapore's knowledge creation. It is consistently ranked as one of the top universities in Asia and is internationally respected for its high-quality research in science, technology and the humanities, and increasingly for research at the interfaces between these areas. The institution includes 16 schools and faculties, and 23 university-level research institutes and centres, focused on critical issues confronting Asia and the world. The university is also home to three of Singapore's five "Research Centres of Excellence" (RCEs), specialising in quantum technologies, cancer and mechanobiology, and is a partner in a fourth national RCE on environmental life sciences and engineering. According to the Nature Publishing Index (NPI) Asia-Pacific 2012, NUS is Singapore's leading research institution. It is ranked ninth in the Asia-Pacific region and is the first Singapore institution to enter the NPI Global Top 100, ranking in 76th position.

In 2013, the NUS placed 22nd in the Times Higher Education "World Reputation" Rankings. In the 2012/2013 QS world rankings by subject, the NUS emerged second among Asian universities for Engineering and technology, Life sciences and medicine, and Natural sciences, and first for Arts, Humanities and Social sciences. Again in the QS rankings, NUS also placed among the world's top 10 universities in 12 disciplines and in the top 20 for a further nine disciplines. The disciplines concerned range from those in the hard sciences (12 disciplines), to others in Geography, Sociology, Modern languages, Accounting & finance, Economics, and English language

Much of the research at NUS is integrated and multi-disciplinary in nature, with particular emphasis on themes such as integrated sustainability solutions for energy, water and the environment; ageing populations; biomedical sciences and translational medicine; global-Asian studies; finance and risk management; and materials science.

In 2012 the NUS had more than 2,270 research-active faculty. Overall, these researchers produced over 7,200 publications and filed roughly 470 patents and 270 invention disclosures. University departments and faculty obtained close to S$580 million in external research funding. The NUS's overall publication of primary research puts it in the top 1 % in 18 out of 22 categories surveyed under the Thomson Reuters Web of Science. NUS faculty serve as consultants and advisors to more than 50 industry and government bodies. Several leading companies have also chosen to establish research labs and partnerships at the NUS, including Siemens, GE, Zeiss and Agilent.

The NUS Kent Ridge campus, which includes the National University Hospital, is located within Singapore's main research district. The university facilities include the new University Town, a study and teaching centre, in turn co-located with the Campus for Research Excellence and Technological Enterprise (CREATE), which brings together top researchers from around the world. Nearby are Biopolis and Fusionopolis, with the A*STAR research institutes and a broad range of public and private labs. The proximity of all these institutions, enterprises and facilities promotes collaboration and synergy between NUS and the broad R&D community, and creates a fertile environment for education, innovation and enterprise.

The NUS is internationally respected for its high-quality research, and the outputs continue to grow rapidly. In 2012, some of the highlights were:

- close to 2,280 "research-active" faculty;
- over 7,200 papers published in internationally refereed journals;

- close to US $425 million in external research grants received;
- over 2,200 on-going projects;
- over 260 research cooperation agreements signed;
- 469 patents filed and 21 licenses granted;
- 276 invention disclosures.

The NUS is actively involved in international academic and research networks such as the Association of Pacific Rim Universities and International Alliance of Research Universities. One of its best-known cooperation ventures is with MIT. In 2007, the two universities created the Singapore-MIT Alliance for Research and Technology, or SMART Centre. SMART is a research and education centre specialising in five areas: Computational engineering; Calculus and systems biology; Manufacturing systems and technology; Advanced materials for micro and nanosystems; Chemical and pharmaceutical engineering.

As was traditional for public universities developed under the Commonwealth tradition, the NUS historically viewed teaching as its primary function, followed by its research role. The 1980s and 1990s saw an increasing emphasis on research, and in the mid-1990s the NUS added the further step of establishing a technology licensing office. The major impetus towards the third university mission came in the late 1990s, when a new vice-chancellor was appointed with strong support from several senior government ministers. Harvard-trained, with a background in American industry (General Electric) and ivy-league administration, the new vice chancellor significantly accelerated the pace of several existing initiatives, and initiated the explicit shift toward the "entrepreneurial university" model (Wong and Singh 2007, 2012). The new entrepreneurial focus of the National University of Singapore favoured local development in the broader sense and contributed directly to the birth of Biopolis.

5.3.1 The Role of the NUS in Development of Biopolis

The opening of Biopolis in October 2003 effectively laid the cornerstone of Phase One of the Singapore Biomedical Sciences (BMS) initiative, following on several years of preliminary work guided by four people: Philip Yeo, Chairman of the Agency for Science, Technology, and Research (A*STAR), Tan Chorh Chuan, then Dean of Medicine at the National University of Singapore (NUS), and oncologists John Wong and Kong Hwai Loong.

However the BMS initiative also represented a refocusing of a decade of previous national and university efforts, in broader-based investment in technology. This previous experience demonstrated the need for greater focus, particularly in the biotechnology sector, and thus opened the road for the creation of Biopolis. The NUS participated directly in the preliminary phases of the BMS through the participation of Tan Chorh Chuan, dean of the medical faculty. NUS also anticipated and stimulated the formation of Biopolis through its long-standing educational role in the sector, producing the necessary experts. As early as 1983, the university had modified its teaching program to focus more closely on biology, and had introduced microbiology as a separate discipline instead of considering it as dependent on biology. As a result, the class hours in microbiology for second and third year students were more than doubled. However changes to the undergraduate curriculum only partly addressed the rising needs in biotechnology, as an emerging, research-intensive field. In fact one year later, a state administrator (Chua Sia Eng, Commissioner of the Parks and Recreation Department) provided a valedictory address for NUS science and pharmacy graduates in which he insisted that the graduates go on to join R&D programs in biotechnology so that Singapore could develop a strong base to tap the vast potential of this science. His call to science and pharmacy graduates to join postgraduate courses was intended to drive home the urgency for qualified biotechnology experts, but in truth seems to have preceded any substantial opening of genuine biotech-centred postgraduate opportunities. This was one of the motivations given for the setting up of the Institute of Molecular and Cell Biology in early 1985. According to NUS's vice-chancellor at the time, the new institute was intended not only to train a pool of competent manpower to service the biotechnology industry, but also offer true research careers to NUS graduates in chemistry and microbiology.

Thus in the early stages preceding Biopolis, the NUS reoriented its traditional activities in education and research specifically towards biology and biotechnology, creating the cultural substrate that would generally favour the birth of a science hub. The National University of Singapore also participated actively in the development of the sector as a whole, collaborating with business, and focusing its research and education in specific areas.

5.3.1.1 Start-up Development

Relative to other universities in the global context, the NUS also moved early to develop its third mission. In 1988 it created the National Entrepreneurship Centre, which in 2001 became part of the NUS Enterprise Start-Up Runway (NUS Enterprise). Both organisations have had the mission of nurturing entrepreneurial learning and venture creation in the NUS community. The NUS also provides advisory services for students to assist them in identifying the paths that will develop the necessary knowledge, skills and contacts for their entrepreneurial initiatives. The NUS Overseas College offers potential entrepreneurs the possibility to broaden their contacts through continuous cooperation and learning in the field, working alongside actors who have created successful start-ups both in Singapore and in other nations. In this way, the NUS offers its student entrepreneurs the possibility of developing not only specialised scientific knowledge, but also the knowledge of business fundamentals necessary for development of their own activities. The NUS has also created Start-Up@Singapore (S@S), an instrument for stimulation and participation in development of territorial start-ups. Start-Up@

Singapore stimulates development of innovative ideas and encourages healthy competition between potential entrepreneurs by organising "global business plan" competitions that offer the winners financing to continue with their innovative activity. As of 2012, the S@S competition was in its 14th edition, representing a total of more than 3,500 teams and 10,000 individual participants, and had helped to spin off more than 90 start-ups. These have included award-winning businesses such as tenCube, FriarTuck, PerceptiveI, World Indigo, PurpleAce and Quantagen. S@S is in fact more than a business planning competition, having evolved into a program that leads students into experiential learning about entrepreneurship, engaging them in the early stages of the actual start-up process: writing a plan, building a team, learning networking skills, pitching to potential investors and getting mentor feedback.

The NUS has also developed more complex actions for development of entrepreneurialism, particularly in the form of NUS Enterprise Start-Up Runway, which is a true enterprise incubator. NUS Enterprise follows the growth process of the new firms, offering tailored services for the different stages of development: pre-incubation, incubation and growth.

In the pre-incubation stage, NUS Enterprise can help start-ups with market validation, user testing and consumer feedback, in order to achieve an optimal product-market fit. This ensures that the potential entrepreneur can build his or her product to cater to their target market, better positioning the new business for take-off. When the business idea is validated and the start-up moves toward the incubation stage, NUS Enterprise offers a wide range of specifically focused support services such as mentorship, talent recruitment and assistance with strategic reviews. It also offers start-ups their own operating spaces, situated in several dedicated infrastructure facilities: NEI@blk71; NEI@pgp and NEI@faculty-of-engineering. In this stage the NUS also provides guided access to its network of relationships, thus opening possibilities for the start-ups to enter markets such as the USA and China, through specific business hubs in territories such as Suzhou, Beijing and Silicon Valley. Start-ups can also leverage NUS incubation services such as mentoring, networking sessions, hot-desking facilities, and overseas grant support, to kick start new expansion stages.

The university also accompanies the start-up firm in the stage of gathering capital investment, supporting relationships with venture capitalists and "angel investors" who have interests in the start-up's areas of activity. NUS also assists start-ups to access grants from the relevant Singapore statutory boards for encouragement of entrepreneurship. The university itself offers a series of grant programs: the Innovation and Entrepreneurship Practicum Grant, Action community for entrepreneurship, Spring Singapore Enabling Enterprise, the First Leap Overseas Grant and the Youth Social Entrepreneurship Programme For Start-Ups.

The NUS's policies and programs for support of entrepreneurial activity have enabled the university to spin off rough 250 start-ups. Many of these have become successful firms in the biology and biotech fields, and thus contribute to the core structure of Biopolis.

Examples of NUS biotech spin-offs

Lynk Biotechnologies

Lynk Biotechnologies is an NUS spin-off incorporated in 2000, for the application of revolutionary technologies in design and development of innovative products, services and applications, for improved quality of life and well-being for all. One of the specificities of Lynk is its scientific planning approach, using SM@RT™ Drug Design, Receptomics™ and PharmaGlue™ technologies, as well as specific knowledge in proteomics. With this approach, Lynk has accelerated and improved efficiencies in the drug discovery process, shortening the lead stages from years to months. One of Lynk's first products was Biolyn™ shampoo, a hair care product based on natural materials, with active ingredients formulated using bio-transformed phyto-therapeutic agents, vitamins and minerals. The product maintains healthy hair and prevents hair loss and thinning. Lynk Biotechnologies is the brainchild of Lee Chee Wee, who developed the technological innovations for the founding of Lynk through his role as a research professor at NUS. Lynk has successfully commercialised these technologies and made them available to the public.

AyoxxA Biosystems

AyoxxA began life in 2008 as a research project to apply parallel biomolecular microarray systems for simultaneous detection of multiple protein biomarkers, in a single step. In the company's words, the innovation represents "Ferrari-level horsepower" in protein analysis, allowing hundreds of diseases, from infectious types to cancer, to be detected from a single blood sample. The new technology is a significant improvement over previous diagnostic technologies, which can detect only a handful of diseases in a single test run. The end results is faster diagnosis, meaning faster treatment for the patient.

The research team, led by Dieter Trau of the faculty of engineering, wanted to develop a product permitting cost-effective manufacturing. Their innovative work led to the incorporation of AyoxxA in 2010, with Professor Trau holding the position of Chief Scientific Officer and another team member, Dr. Andreas Schmidt, serving as CEO. The spin-off company is developing NUS's proprietary technology as a platform to allow simultaneous, precise and cost-effective protein analysis from minute samples.

Clearbridge BioMedics, NanoMedics and VitalSigns

Johnson Chen is a "serial entrepreneur", having launched three healthcare start-ups based on technologies licensed from the NUS. The series of company set-ups started when Mr. Chen met up with Lim Chwee Teck, a friend from past years at Cambridge, then a professor in the NUS departments of mechanical engineering and bioengineering, who was developing a technology that trapped tumour cells circulating in blood. Realising the commercial potential of the technology, the two friends spun it off as Clearbridge BioMedics in 2009. They translated the lab prototype into a commercial product and launched it as the ClearCell™ system, which detects, isolates and retrieves intact circulating tumour cells. As of 2011, ClearCell™ was involved in global marketing to the research community, while the company turned its focus to developing a new generation of the system for clinical applications.

Recognising that the NUS held further technologies with commercial interest, Mr. Chen then licensed another of Prof Lim's innovations: a bioresorbable and biocompatible nanofibre with good mechanical, physical and chemical properties. Clearbridge NanoMedics was set up to commercialise the technology for cosmetic and injury management applications. NUS then also introduced Johnson to an ultra-low powered electrocardiogram (ECG) chip, invented by Lian Yong, a professor of electrical and computer engineering. This let to establishment of Clearbridge VitalSigns, which is developing the NUS technology as CardioLeaf®, a fully-integrated 3-Lead ECG monitor, targeting both medical and fitness applications.

Source NUS Enterprise organisational website

5.3.1.2 Patents and Licensing

As of 2013 the National University of Singapore had registered over 3,500 patents and issued 250 technology licenses. Searches in the World International Patent Office records show that the university holds 307 patents in the field of Medical and veterinary science and 207 in Biochemistry.

Major NUS license agreements under Biotech

Factor C

"Factor C" is a technology first developed in NUS labs that is now being applied globally in the development of safer injectable medications, such as vaccines and intravenous drugs. The same technology is also saving the lives of the horseshoe crab, an endangered species in many of its global habitats. The husband-wife team of professors Ho Bow and Ding Jeak Ling, respectively from the departments of microbiology and biological sciences, developed the original research, which has become one of the NUS's most successfully commercialised technologies.

Pharmaceutical and medical device companies routinely use horseshoe-crab blood to ensure medication and equipment are free from bacterial endotoxins. This is possible due to an enzyme in the crab's blood, Factor C, which triggers the clotting process when in contact with bacterial endotoxins.

However collecting horseshoe crab blood is a time-consuming and costly process. Professors Ding and Ho developed a method for laboratory production of Factor C, using a recombinant DNA-based system to clone the original enzyme. Not only does the clone succeed in reacting with endotoxins, and at very low concentrations, it also results as more stable and chemically-consistent than the natural form. With the support of the NUS Industry Liaison Office, this ground-breaking technology was licensed for production and introduced for global use. Lonza, the international life-sciences leader, has since incorporated the NUS technology into its endotoxin detection kits and online endotoxin monitoring systems. Another US-based company, BioDtech Inc., has applied related technologies in development of products for removal of bacterial endotoxins from pharmaceuticals and fluids.

Rapid-assay diagnostic kits for detection of parasitic diseases
Malaria and dengue fever are mosquito-borne tropical diseases that affect millions of people, while malaria kills about three million people per year. Rapid, accurate diagnosis is paramount for timely emergency response, treatment and containment. The standard test for the malarial *plasmodium* parasite is time-consuming, laborious, and can produce false negatives. Testing for dengue fever takes up to 8 days and can also deliver inaccurate results. However a medical-diagnostics company, using breakthrough molecular technology from the NUS, has developed rapid-assay test kits that detects the parasites in a matter of hours.

NUS researchers Ursula Kara, Robert Ting, Jill Tham, James Nelson and Theresa Tan discovered and patented the unique nucleic acid diagnostic primers for the parasitic vectors over a ten-year period, finally announcing the technology in 1998. A primer is a short strand of DNA/RNA that serves in the formation of longer strands. Using a single drop of blood, the highly sensitive polymerase chain-reaction technology can distinguish between different *plasmodium* species within 3 h. The dengue fever kit can detect the virus within three to 5 days after it first appears in the bloodstream, compared to the usual 8 days for standard immunodiagnostic methods. Detection enables earlier medical attention, which can be critical for preventing serious complications such as dengue haemorrhagic fever or shock syndrome. The NUS has licensed the technology to Veredus Laboratories, which is manufacturing and selling several different diagnostic kits. Singapore's National University Hospital has used the dengue fever kit for more than 3 years as a routine diagnostic tool. In addition, Veredus has produced the world's first validated commercial avian flu diagnostic kit, which cuts the time for accurate detection of the H5N1 virus from 7 days to as little as two. As of 2013, the company was also developing kits for encephalitis, Japanese encephalitis, yellow fever, chicken pox and SARS.

Anti-restenosis drug

Restenosis is a recurring occurrence of narrowing of the arteries. A new drug with a novel mechanism for preventing the condition was first discovered by Sim Meng Kwoon, an assistant professor in the NUS department of pharmacology. The drug des-aspartate-angiotensin I (DAA-I) acts on a receptor coupled to a specific prostaglandin pathway to suppress clot formation. The highly specific pathway limits any negative secondary effects. Currently, metal stents that are inserted into patients' blood vessels after angioplasty are coated with a bio-degradable polymer impregnated with drugs to prevent restenosis. However the existing stent drugs like Sirolimus and Paclitaxel are nonspecific immunosuppressant and anti-cancer drugs, which can also affect other cells.

In commercialising the technology, NUS has entered into a licensing agreement with Medlogics Device Corporation, a US start-up company founded by a group of professionals with experience in medical devices and stents. The agreement was the first ever case where a Singapore drug innovation will enter global commercialised through the United States. It is also the first ever case where a specific new anti-restenosis drug is being incorporated into an equally new Medlogics technology. Medlogics has a proprietary coating technology that enables the drugs to be applied directly to the stent metal, avoiding the use of biodegradable polymers. The technology reduces the cost of making the drug-coated stents, ensures consistent release of drugs, and reduces the inflammation caused by the polymers. Medlogics selected DAA-I as the first drug for application with the new technology, out of a short list of seven final candidates.

Similar to the other cases cited here, the NUS Industry Liaison Office played a key role in flanking Dr. Sim during the development and negotiation of the exclusive license agreement between the university and Medlogics.

Source NUS Industrial Liaison Office

5.3.1.3 University-Industry Collaborations

The National University of Singapore is party to numerous agreements with major international biotech companies, such as Bayer Healthcare, Merck and GlaxoSmithKline. The trend towards these agreements is increasing, with 35 % growth between 2011 and 2013.

The collaborations between the NUS and private firms are of different kinds and cover a wide range of activities. In some cases they are true cooperative research projects drawing in widespread research teams of top professionals, such as in the case of current work towards a cure for diabetes using adult stem cells.

Stem cell research project

This research project, for the creation of insulin-secreting cells from the patient's own adult stem cells, unites a group of global leaders in basic research, clinical applications, industry collaboration and surgery. The main research focus is on the use of adipose tissue, but other sources of adult stem cells may also be investigated. The research team includes Bernat Soria, director and professor in physiology for the Alicante Institute of Bioengineering (Spain) presently an adjunct professor with the NUS department of surgery, and Sir Roy Calne, Professor of Surgery Emeritus at Cambridge. Dr. Calne was knighted in 1996 for his pioneering work in organ transplantation. The team also includes John Isaac, acting head and education director for the NUS department of surgery, head of the hepato-biliary pancreatic surgery division, and senior consultant to the NUS liver transplantation service. Finally, the team also includes Dr. Susan Lim, founder, chairman and CEO of Stemcell Technologies Pte. Ltd., an accomplished general surgeon in active surgical practice, known for her pioneering work in liver transplants and for performing the first transplants in the Asian region.

Source NUS website

In other cases the collaborations arise from funding agreements and grants by private firms, for research towards the solution of specific problems. For the NUS, there are numerous examples of this type of arrangement in the pharmaceutical and biotech sector.

Major NUS-firm project collaborations

Bayer HealthCare Grants4Leads

Bayer HealthCare's research and development focuses on identifying and developing new active substances for treating diseases with high unmet medical needs. The NUS collaboration initiative seeks to identify and develop novel small-molecule leads, particularly in the fields of oncology, cardiology, haematology and gynaecology therapies.

Bayer HealthCare Grants4Targets

Bayer HealthCare provides grants to the NUS for the exploration of attractive, novel drug targets in the fields of oncology, gynaecology, cardiology, and haematology, with the intent to accelerate the transition from basic research towards promising new treatment options.

Eli Lily Open Innovation Drug Discovery
Eli Lilly has established a program to provide NUS academic researchers access to proprietary, disease-relevant phenotypic assays (PD2) and sophisticated in vitro target-based assays (TargetD2) for the screening of small molecules targeting cancer, neurological disorders, and metabolic diseases.

GlaxoSmithKline Academic Centre of Excellence Singapore
The Academic Centre of Excellence is a virtual research network for projects of mutual interest to GlaxoSmithKline and Singapore academic scientists, as well as for development of partnerships to accelerate translational research and progress toward shared goal in providing accessibility to new medicines.

LIMR Chemical Genomics Centre
The LIMR Chemical Genomics Centre connects university and pharmaceutical scientists in a totally new public-private translational-research framework for testing the "druggability" of innovative cell-signalling targets, called Double-Blinded Drug Discovery® (DBD2).

Merck Investigator Studies Programme
This Merck-NUS joint program is designed to advance science and improve patient care by supporting the provision of drugs and vaccines, through total or partial funding of high-quality research that is initiated, designed, implemented and sponsored by external investigators.

Source NUS Annual Report

In other cases the university-firm relationships take the form of private financing for entire research centre programmes. In 2012–2013, the NUS received roughly 184 million US$ in this manner, 31 % of which was directed to the Yong Loo Lin School of Medicine, 12 % to the Duke-NUS Graduate Medical School and 8 % to the Faculty of Science.

5.3.1.4 Education and Research

As noted previously, the NUS maintains a long tradition of education and research in the fields of biology, medicine and biotechnology, which has produced many of the qualified personnel involved in the birth and growth of Biopolis. Currently the NUS has 16 faculties and schools distributed over three campuses in the Singapore city-state: Kent Ridge, Bukit Timah and Outram. In terms of the growth of Biopolis, the most important of these are the undergraduate schools in medicine and science and the advanced-degree programs in medicine and science, including those offered by the Duke-NUS Graduate Medical School and the Institute of Systems Science, which are directly linked to Biopolis. In 2013 alone, the various faculties produced 1,637 graduates and 1,345 biological and medical sciences specialists.

In addition to carrying out its education function, the NUS has 23 research centres, including the Life Sciences Institute and the Singapore Institute for Neurotechnologies, both of which are specifically linked to Biopolis. The NUS also has four global-level "centres of excellence", two of which are at the core of the Biopolis cluster: the Cancer Institute of Singapore and the Mechanobiology Institute of Singapore.

Biotech "centres of excellence"

Cancer Science Institute of Singapore

The Cancer Science Institute was launched in October 2008, with the aim of positioning Singapore as a global leader in biomedical sciences. Its mission is to conduct a multifaceted and coordinated approach to cancer research, extending from basic cancer studies to experimental therapeutics. CSI Singapore is a state-of-the-art university research institute affiliated with and hosted by the National University of Singapore. The institute was founded on the basis of a $172 million "Research Centre of Excellence" grant, one of only five allocated by the Singapore National Research Foundation and the Ministry of Education.

The institute is an anchor for research expertise in two broad programs: Cancer biology and stem cells, and Experimental therapeutics. These are platforms for CSI Singapore's approaches to key research themes in leukaemia and cancer diseases of the liver and lungs, which are endemic in Asian populations. In addition, the institute has also recently started a working group on breast cancer, looking at Asian phenotypes. CSI Singapore is housed on three floors of the Centre for Translational Medicine, within the Yong Loo Lin School of Medicine. The location is opposite the thousand-bed National University Hospital and close to a main subway line providing a link to the A*Star research institutes at Biopolis, Duke-NUS Graduate Medical School and a second major hospital, the Singapore General. Core facilities and support technologies include the Centre for Translational Research and Diagnostics, a fluorescence-activated cell sorting facility, the Bioinformatics Core, a leukaemia cell bank, a transgenic and gene targeting facility and the Xenograft Cancer Models Facility. A number of the researchers working with CSI Singapore also hold positions at renowned international institutions such as the Beth Israel Deaconess Medical Center, Harvard Medical School, Brigham and Women's Hospital of Boston, Sweden's Karolinska Institute, the École Normale Supérieure and Johns Hopkins University.

Mechanobiology Institute of Singapore

The Mechanobiology Institute was founded in 2009, under joint funding from the National Research Foundation and the Ministry of Education. The aim of the institute is both to advance the discipline and benefit the Singapore population. MBI's primary research focus is the identification, measurement and description of the forces for motility and morphogenesis at the molecular, cellular and tissue level. Toward that goal, the institute is working to create a common international standard for defining these steps by developing powerful new computational models, experimental reagents, and tools for studying diseases of cells and tissues. These basic discoveries are then transferred to both the clinic and the classroom, as a basis and support for scientific progress in the field of Mechanobiology. Towards these goals, the MBI has developed memoranda of understanding with a range of organisations, promoting cross-talk and active collaboration between MBI and partner institutions such as France's Centre National de la Récherche Scientifique (CNRS), the Indian Institute of Science Education and Research, the National Centre for Biological Sciences in Bangalore (India), and the Waseda University in Japan. The focal areas for MBI research area: (i) stem cells and tissue engineering, (ii) development and cancer, (iii) microbes and pathogenesis.

Source NUS website

In terms of bibliographic indicators of research production, the NUS produces roughly 7,500 published papers per year and is ranked by Thomson Reuters Web of Science in the top 1 % of global institutions for 18 out of 22 categories surveyed. A search of Scopus Database shows that NUS researchers and affiliated research centres have produced 2,172 papers in the fields of medical sciences and biology.

In conclusion, we identify that the NUS serves as an important driver in the development of the Biopolis LIS, through:

- education and training of specialised personnel;
- high-profile research in the fields of biology and medicine;
- cooperation with local and foreign firms;
- creation of a system of innovative firms.

These NUS initiatives support the growth of the system by stimulating the direct and indirect creation of a network of local firms alongside the multinational firms present on the territory. The NUS initiatives favour the university-firm relations, and produce the binding effects that permit the system to transform into a true LIS.

References

Altbach P (1998) Comparative higher education: knowledge, the university, and development. Ablex, Greenwich
Bernardes A, Albuquerque E (2003) Cross-over, thresholds, and interactions between science and technology: lessons for less developed countries. Res Policy 32:865–885
Cassia L, Colombelli A, Paleari S (2008) Regional transformation processes through the universities-institutions-industry relationship. Ind High Educ 22(2):105–118
Chang YC, Yang PY, Chen MH (2009) The determinants of academic research commercial performance: towards an organizational ambidexterity perspective. Res Policy 38:936–946
Charles D (2006) Universities as key knowledge infrastructures in regional innovation systems. Innov Eur J Soc Sci 19(1):117–130
Chatterton P, Goddard J (2000) The response of higher education institutions to regional needs. Eur J Educ 35(4):475–496
Cooke P (2002) Regional innovation systems: general findings and some new evidence from biotechnology clusters. J Technol Trans 27(1):133–145
Di Gregorio D, Shane S (2003) Why do some universities generate more start-ups than others? Res Policy 32:209–227
Etzkowitz H (2002) MIT and the rise of entrepreneurial science. Routledge, London
Etzkowitz H, Leydesdorff L (1997) Introduction: universities in the global knowledge economy. In: Etzkowitz H, Leydesdorff L (eds) Universities and the global knowledge economy: a triple helix of university-industry government relations. Pinter, London, pp 1–8
Etzkowitz H, Leydesdorff L (1999) The future location of research and technology transfer. J Technol Transf 24:111–123
Etzkowitz H, Webster A, Gebhardt C, Terra B (2000) The future of the university and the university of the future: evolution of ivory tower to entrepreneurial paradigm. Res Policy 29:313–330
Freeman C (1995) The national system of innovation in historical perspective. Camb J Econ 19:5–24
Freeman C (1991) Networks of innovators: a synthesis of research issues. Res Policy 20(5):499–514
Fritsch M, Schwirten C (1998) Öffentliche forschungseinrichtungen im regionalen innovationssystem. Raumforschung und Raumordnung 4(98):288–298
Gunasekara C (2006) Reframing the role of universities in the development of regional innovation systems. J Technol Transf 31(1):101–113
Guston DH (2000) Retiring the social contract for science. Issues Sci Technol 16(4):32–38
Hall P (1994) Innovation, economics and evolution. Harvester Sheaf, New York
Hart DM (1988) Forged consensus: science, technology and economic policy in the United States, 1921–1953. Princeton University, Princeton
Holland BA (2001) Toward a definition and characterization of the engaged university. Metrop Univ 2(3):20–29
Lee YS, Tee YC (2009) Reprising the role of the developmental state in cluster development: the biomedical industry in Singapore. Singap J Trop Geogr 30(1):86–97
Liefner I, Hennemann S, Xin L (2006) Cooperation in the innovation process in developing countries: empirical evidence from Zhongguancun, Beijing. Environ Plann A 38(1):111–130
Lundvall B (1992) Introduction. In: Lundvall B (ed) National systems of innovation: towards a theory of innovation and interactive learning. Pinter, London
Lundvall BÄ., Johnson B (1994) The learning economy. J Indus Stud 1(2):23–42
MTI (2006) Sustaining innovation-driven growth: science and technology 2010 plan. Ministry of Trade and Industry, Singapore
Murray F (2004) The role of academic inventors in entrepreneurial firms: sharing the laboratory life. Res Policy 3:643–659
Niosi J, Bas TG (2001) The competencies of regions–Canada's clusters in biotechnology. Small Bus Econo 17(1–2):31–42

O'Shea RP, Chugh H, Allen TJ (2008) Determinants and consequences of university spinoff activity: a conceptual framework. J Technol Transf 33:653–666
OECD (1999) The response of higher education institutions to regional needs. Centre for Educational Research and Innovation, Organisation for Economic Cooperation and Development, Paris
OECD (2007) Higher education and regions: globally competitive, locally engaged. Organisation for Economic Cooperation and Development, Paris
OECD (2013) Innovation in Southeast Asia. OECD Reviews of Innovation Policy, Organisation for Economic Cooperation and Development, Paris
Perez C, Soete L (1988) Catching up in technology: entry barriers and windows of opportunity. In: Dosi G, Freeman C, Nelson R, Silverberg G, Soete L (eds) Technical change and economic theory. Pinter, London, pp 458–479
Philpott K, Dooley L, O'Reilly C, Lupton G (2011) The entrepreneurial university: examining the underlying academic tensions. Technovation 31(4):161–170
Power D, Malmberg A (2008) The contribution of universities to innovation and economic development: in what sense a regional problem? Camb J Reg Econ Soc 1(2):233–245
Powers J (2004) R&D funding sources and university technology transfer: what is stimulating universities to be more entrepreneurial? Res High Educ 45(1):1–23
Salter B. (2007) State strategies and speculative innovation in regenerative medicine: the global politics of uncertain futures. Global biopolitics research group working paper 20
Schiller D (2006a) The potential to upgrade the Thai innovation system by university-industry linkages. Asian J Technol Innov 14(2):67–91
Schiller D (2006b) Nascent innovation systems in developing countries: university responses to regional needs in Thailand. Ind Innov 13(4):481–504
Schiller D, Liefner I (2007) Higher education funding reform and university: industry links in developing countries—the case of Thailand. High Educ 54(4):543–556
Shane S (2004a) Academic entrepreneurship: university spin-offs and wealth creation. Edward Elgar, Cheltenham
Shane S (2004b) Encouraging university entrepreneurship? The effect of the Bayh-Dole act on university patenting in the United States. J Bus Ventur 19(1):127–151
Smith BLR (1990) American science policy since World War II. Brookings Institution, Washington DC
Sutz J (1997) The new role of the university in the productive sector. In: Etzkowitz H, Leydesdorff L (eds) Universities and the global knowledge economy: a triple helix of university-industry-government relations. Pinter, London, pp 11–20
Uyarra E (2010) Conceptualizing the regional roles of universities: implications and contradictions. Eur Plan Stud 18(8):1227–1246
Van Looy B, Ranga M, Callaert J, Debackere K, Zimmermann E (2004) Combining entrepreneurial and scientific performance in academia: towards a compounded and reciprocal Matthew-effect? Res Policy 33:425–441
Wong KW, Bunnell T (2006) 'New economy' discourse and spaces in Singapore: a case study of one-north. Environ Plann A 38(1):69–83
Wong PK, Singh A (2005) The national system of innovation in Singapore. In: Edquist C, Hommen L (eds) Small country innovation systems: globalization, change and policy in Europe and Asia. Elgar, Cheltenham
Wong PK (2006a) Toward an ecosystem for innovation in a newly industrialized economy: Singapore and the life sciences. Ind High Educ 20(4):231–236
Wong PK (2006b) The re-making of Singapore's high tech enterprise ecosystem. In: Rowen H, Miller W, Hancock M (eds) Making IT: the rise of Asia in high tech. Stanford University, Palo Alto, pp 123–174
Wong PK (2011) The dynamism of Singapore's science and technology policy and its quest for technopreneurship. In: Mian SA (ed) Science and technology based regional entrepreneurship: global experience in policy and program development. Edward Elgar, Cheltenham

Wong PK, Ho YP, Singh A (2007) Towards an "entrepreneurial university" model to support knowledge-based economic development: the case of the National University of Singapore. World Dev 35(6):941–958
Wong PK, Ho YP, Singh A (2012) Towards a 'global knowledge enterprise': the entrepreneurial university model of the National University of Singapore. In: Wong PK (ed) Academic entrepreneurship in Asia: the role and impact of universities in national innovation systems. Edward Elgar, Cheltenham, pp 165–198
World Bank (2000) Higher education in developing countries. World Bank, Washington
Wright M, Lockett A, Clarysse B, Binks M (2006) University spin-out companies and venture capital. Res Policy 35:481–501
Yue CS, Lim JJ (2002) Singapore: a regional hub in ICT. In: Masuyama S, Vandenbrink D (eds) Towards a knowledge-based economy: East Asia's changing industrial geography. Institute of Southeast Asian Studies, Singapore, pp 259–298

Chapter 6
Conclusions

Through recent analyses of a number of single systems, scholars have demonstrated that the birth of innovation systems in developing countries occurs through processes that are different from those in developing countries. Innovation systems in emerging nations have been noted as presenting a range of characteristic features, in particular: (i) new knowledge generally originates from outside the system, often from the actions of foreign firms; (ii) multinationals are present and serve as an important external element in stimulating the birth of local systems; (iii) the local research component is limited, and finally, (iv) government institutions assume key roles.

The observation of these specificities has given rise to extensive debate over the validity and universality of the traditional understanding of innovation systems. In spite of this debate, the existing contributions do not seem to propose an alternative to the classic theoretical model, which would describe what occurs in developing countries.

The observations concerning developing nations have in fact generated strong scepticism about the application of the LIS concept to such cases. A fundamental concern among scholars is that the concept of innovation could itself be inapplicable, because developing countries are not capable of realising innovation in the true sense.

Another doubt is that the systemic aspects of an LIS could function, because in the developing countries there is frequently an insufficient basis of useful relationships between the different actors. As a consequence, some authors affirm that the LIS concept can only be applied to developing countries if it is conceived in an evolutionary perspective (Arocena and Sutz 2000; Chaminade and Vang 2008). Reasoning similarly, the specificity of LISs in developing countries need not be considered as a failure to achieve traditional models of innovation systems, and should instead stimulate a reinterpretation of the model in a dynamic mode (Lundvall 2007).

To better understand the dynamics of innovation development in emerging nations, in this volume we have outlined a sort of evolutionary route, in which the end result of an LIS, as it is structured in developed nations, is in fact the point of arrival.

M. Ferretti and A. Parmentola, *The Creation of Local Innovation Systems in Emerging Countries*, SpringerBriefs in Regional Science,
DOI 10.1007/978-3-319-10440-9_6

Based largely on the existing literature, we have theorised a sort of ideal path for the constitution of innovation systems, depending on the actors involved and the developmental level of the local entrepreneurial system.

In the case that the development of the entrepreneurial system is at a very low level and lacks the actors that can participate in the creation of an LIS, the only means to engage such creation mechanisms lies in strong government intervention. This implies appropriate policy, aimed above all at attracting the actors lacking in the territory and favouring their geographic concentration. Government in this case assumes the roles of both the sole driver of the system and key actor in guiding the process of creating an LIS, following a top-down approach.

In the case that the local entrepreneurial systems are more developed and a large company already exists in situ, it could be that this firm, pushed by the necessity of broadening its competitive horizons, would engage virtuous mechanisms for the creation of a LIS. When large companies localise in a specific territory they create a range of strong and weak links with other local actors, which can contribute to the growth of the system. The large firms can interact with the other local firms, above all instituting supply relationships, but also establishing mechanisms for knowledge transfer through cooperative ventures and relations. Large firms can also interact with research centres, financing research or training programmes and again establishing other cooperation mechanism. The large firms can even create other actors in the system through budding processes and stimulation of spin-off companies, and through creation of true corporate research centres and corporate universities.

In other emerging-economy cases the motor for creation of an LIS can be a university, which broadens its mission from that of a research and training institution to that of an actor that guides entrepreneurial development in an area. Enactment of the so-called "third mission" of the university can engage processes of entrepreneurial growth through the creation of academic spin-offs. Spinning off the innovation permits the university to act directly in implementing local entrepreneurial systems. Concession of patents under license or direct commercialisation of research results can also provide a push towards revitalising local firms, and an opportunity to undertake activity with high innovation content. The university can contribute to the birth of the system through individual cooperation agreements, guiding the firm in the difficult phase of research.

In summary, the structuring of an LIS as typically defined for the developed states is in fact a "point of arrival". In the developmental nations the entrepreneurial system may lack one or more key actors, and hence require engagement of prior creation processes that in time can lead to the creation of an LIS. In this sense the LIS concept is applicable in emerging nations only if the key actors share the conception of the dynamic nature of the LIS, as the final point in a process of creation.

In this regard we have hypothesised the existence of three different configurations of LISs in emerging nations, representing diverse intermediate stages, all in the process of convergence towards the configuration of an LIS as traditionally understood.

The emergence of the different configurations depends on the degree of development of the local entrepreneurial system and foresees the existence of a specific actor capable of engaging the creation process. In this sense, according to the actor involved, we distinguish the three configurations of LISs:

- the government-driven LIS;
- the firm-driven LIS;
- the university-driven LIS.

The government-driven LIS develops in situations where the local entrepreneurial system has a development level such that there is no other actor capable of intervening in the process of system creation. Firm-driven and university-driven LISs take hold in situations where the entrepreneurial system is more advanced.

All three configurations will succeed over time if the guiding actor is capable of engaging positive mechanisms for involvement of other actors in the creation and construction of a true LIS.

As the chapters in this volume have proceeded, we have explored the various aspects of this model in further detail, also providing relevant case studies. In Fig. 6.1 we provide a further classification of the case studies presented in the volume, according to the model.

The nations of Iran and the United Arab Emirates (Dubai) offer examples of attempts to construct LISs in situations where the local entrepreneurial system presents a low level of development. We identify this level on the basis of the numbers of local firms operating in sectors that are considered "high-tech", and not necessarily on the basis of the overall economic development. In fact if we consider these two nations, the general economic development is very high, but has thus far not included significant numbers of firms operating in high-tech sectors. Both cases represent examples where the government has intervened to create local entrepreneurial systems through mechanisms that attract foreign firms and favour the development of relations with local ones. The Iranian and Dubai cases illustrate two different policies that governments can adopt, involving respectively direct or indirect intervention, both of which can have positive long-term results.

Fig. 6.1 Classification of case studies in the current volume

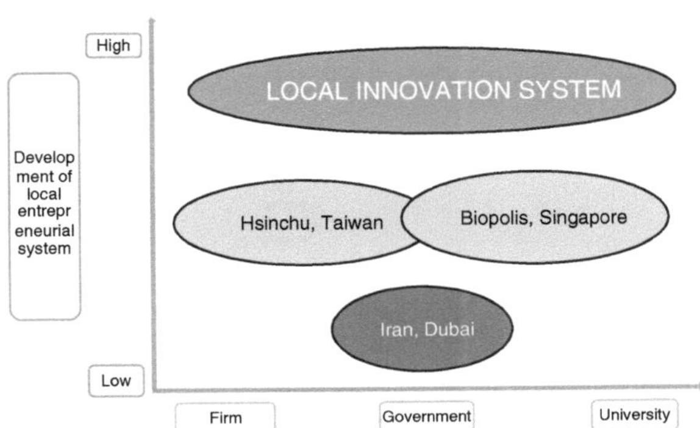

In cases of more developed entrepreneurial systems it can be a local firm, as with Acer Inc. in Taiwan, or a university, as with the National University of Singapore, that activates the various mechanisms to connect the other actors necessary for the birth of an LIS.

The case studies examined in the volume are useful because they bring out the dynamic nature of the model, under which the different systems evolve and ultimately converge on the state of a true LIS. For example in the case of government-driven LISs, the strategies enacted by the local governments are in some cases creating the conditions for transformation into a firm-driven LIS. For the AID-Renault case, in Iran, the relationship with the foreign firm has created an improvement in the local firm's knowledge basis, enabling it to compete on local and international markets, and thus in turn to transfer knowledge to other local actors. In the case of Dubai Biotech, the localization of foreign high-tech firms in a defined area has favoured the birth of local firms operating in the biotech sector, which can in future serve as actors that guide the birth of a new LIS. The cases of Acer, in Taiwan, and the National University of Singapore, instead demonstrate how a single actor in the system—whether firm or university—can begin to activate a series of mechanisms for the creation of missing elements in the helix, which interact to give life to a new LIS. In Fig. 6.2 we summarise the evolutionary dynamics active in the development of emerging-nations LISs.

The current volume offers useful theoretical and empirical results. From a theoretical point of view, the book is relevant to the IS literature, since compared to existing contributions that are limited to ex-post analysis of the characteristics and peculiarities of specific systems of innovation, it provides a key to reading that permits examination of the phenomenon of birth of innovation systems in developing nations in a dynamic mode, adopting an ex-ante perspective. The distinction between government-driven, firm-driven and university-driven LISs permits the actors in different nations to identify the mechanisms that the various actors can engage, according to the specific case, for the creation of LISs. Finally, the use of the case study approach permits the examination of emerging-economy LISs in a new perspective, compared to what has until now been seen in the literature

Fig. 6.2 Evolution of LIS

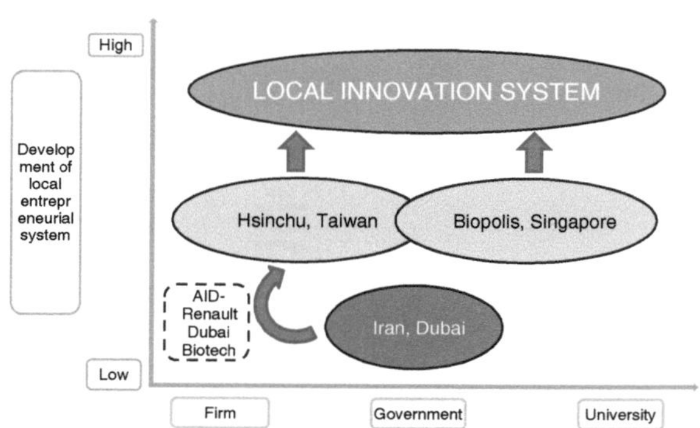

(reports on the Acer and NUS cases) or to deepen the knowledge of cases that have not previously been examined (examples of Dubai Biotech and Iran).

The proposed model offers a useful key for readings by policy-makers, who can define policies for the construction of innovation systems in developing nations that take on a dynamic perspective, identifying interventions formulated specifically for the various phases of development.

The current study opens numerous avenues for further research. Among these, further analysis of innovation systems in developing countries is needed to verify the applicability of the theoretical model in other geographical contexts. As assistance for the policy-maker, further research could better identify the various potential strategies to different support the stages and cases of IS development.

References

Arocena R, Sutz J (2000) Looking at national systems of innovation from the South. Indus Innov 7(1):55–75

Chaminade C, Vang J (2008) Globalisation of knowledge production and regional innovation policy: supporting specialized hubs in the Bangalore software industry. Res Policy 37:1684–1696

Lundvall BA (2007) National innovation systems—analytical concept and development tool. Indus Innov 14(1):95–119

MIX
Papier aus verantwortungsvollen Quellen
Paper from responsible sources
FSC® C105338

If you have any concerns about our products,
you can contact us on
ProductSafety@springernature.com

In case Publisher is established outside the EU,
the EU authorized representative is:
Springer Nature Customer Service Center GmbH
Europaplatz 3, 69115 Heidelberg, Germany

Printed by Libri Plureos GmbH
in Hamburg, Germany